# MAP READING AND LOCAL STUDIES
## IN COLOUR

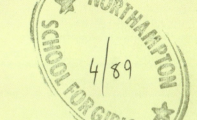

**A. P. Fullagar** B.A.
*Head of Geography Department,
The John Cleveland College,
Hinckley, Leicestershire*

**H. E. Virgo** M.A.
*Late Vice Principal,
The John Cleveland College,
Hinckley, Leicestershire*

---

*British Library Cataloguing in Publication Data*

Fullagar, A. P.
   Map reading and local studies in colour.—3rd ed
   1. Maps   2. Great Britain—Description
and travel—1971–
   I. Title   II. Virgo, H. E.
   912'.01'4   GA151

ISBN 0 340 35144 6
First edition 1967
Second edition 1975
Third edition 1984
Third impression 1988

Printed in Hong Kong for
Hodder and Stoughton Educational,
a division of Hodder and Stoughton Ltd,
Mill Road, Dunton Green, Sevenoaks, Kent TN13 2YD,
by Colorcraft. Photoset by Rowland Phototypesetting Ltd
Bury St Edmunds, Suffolk

Copyright © 1967, 1975, 1984
A. P. Fullagar and H. E. Virgo
All rights reserved. No part of this publication may be
reproduced or transmitted in any form or by any means,
electronic or mechanical, including photocopy, recording or
any information storage and retrieval system, without the
permission in writing from the publisher.

**HODDER AND STOUGHTON**
LONDON    SYDNEY    AUCKLAND    TORONTO

# ORDNANCE SURVEY SYMBOL CHARTS

## Key for 1:50 000 maps

### ROADS AND PATHS
Not necessarily rights of way

- Motorway (dual carriageway)
- Motorway under construction
- Trunk road
- Main road
- Main road under construction
- Secondary road
- Narrow road with passing places
- Road generally more than 4 m wide
- Road generally less than 4 m wide
- Other road, drive or track
- Path
- Gradient: 1 in 5 and steeper    1 in 7 to 1 in 5
- Gates    Road tunnel
- Ferry (passenger)    Ferry (vehicle)

### PUBLIC RIGHTS OF WAY
(Not applicable to Scotland)

- Footpath
- Bridleway
- Road used as a public path
- Byway open to all traffic

Public rights of way indicated by these symbols have been derived from Definitive Maps as amended by later enactments or instruments held by Ordnance Survey on 1st June 1983 and are shown subject to the limitations imposed by the scale of mapping

**The representation on this map of any other road, track or path is no evidence of the existence of a right of way**

Danger Area    MOD Ranges in the area.    Danger! Observe warning notices

### RAILWAYS

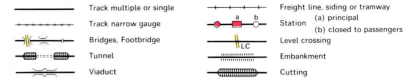

- Track multiple or single
- Track narrow gauge
- Bridges, Footbridge
- Tunnel
- Viaduct
- Freight line, siding or tramway
- Station (a) principal (b) closed to passengers
- Level crossing
- Embankment
- Cutting

### WATER FEATURES

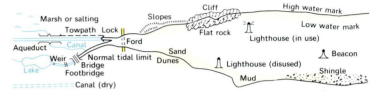

### GENERAL FEATURES

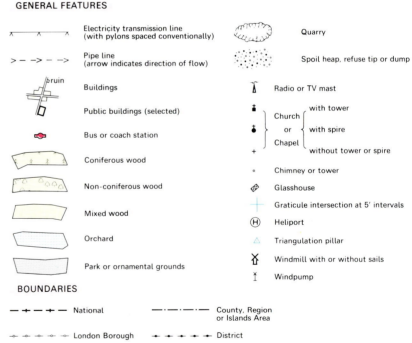

- Electricity transmission line (with pylons spaced conventionally)
- Pipe line (arrow indicates direction of flow)
- Buildings
- Public buildings (selected)
- Bus or coach station
- Coniferous wood
- Non-coniferous wood
- Mixed wood
- Orchard
- Park or ornamental grounds
- Quarry
- Spoil heap, refuse tip or dump
- Radio or TV mast
- Church or Chapel { with tower / with spire / without tower or spire }
- Chimney or tower
- Glasshouse
- Graticule intersection at 5' intervals
- Heliport
- Triangulation pillar
- Windmill with or without sails
- Windpump

### BOUNDARIES

- National
- London Borough
- National Park or Forest Park
- National Trust
- Forestry Commission
- County, Region or Islands Area
- District
- NT always open
- NT opening restricted
- Pedestrians only -observe local signs

ABBREVIATIONS

| | | | |
|---|---|---|---|
| P | Post office | CH | Clubhouse |
| PH | Public house | PC | Public convenience (in rural areas) |
| MS | Milestone | TH | Town Hall, Guildhall or equivalent |
| MP | Milepost | CG | Coastguard |

ANTIQUITIES

VILLA  Roman          ⚔ Battlefield (with date)       ✛ Position of antiquity which cannot be drawn to scale
Castle Non-Roman      ✳ Tumulus

𝔐 Ancient Monuments and Historic Buildings in the care of the Secretaries of State for the Environment, for Scotland and for Wales and that are open to the public

The revision date of archaeological information varies over the sheet

HEIGHTS

Contours are at 10 metres vertical interval

•144  Heights are to the nearest metre above mean sea level

ROCK FEATURES

outcrop
cliff
scree

Heights shown close to a triangulation pillar refer to the station height at ground level and not necessarily to the summit.

On pages 19, 31, 36, 51 and 63 the contours are at 50 feet vertical interval.

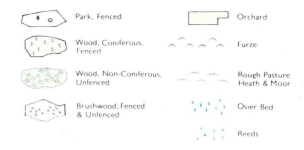

VEGETATION

Park, Fenced — Orchard
Wood, Coniferous, Fenced — Furze
Wood, Non-Coniferous, Unfenced — Rough Pasture Heath & Moor
Brushwood, Fenced & Unfenced — Osier Bed
— Reeds

# Key for 1:25 000 maps

ROADS AND PATHS

Not necessarily rights of way

| M 1 or A 6(M) | Motorway |
| A 31(T) | Trunk road |
| A 35 | Main road |
| B 3074 | Secondary road |
| A 35 | Dual carriageway |
| | Road generally more than 4m wide |
| | Road generally less than 4m wide |
| | Other road, drive or track |

Unfenced roads and tracks are shown by pecked lines

Path

PUBLIC RIGHTS OF WAY

Public paths { Footpath / Bridleway }
Road used as a public path

BOUNDARIES

— · — · — County (England and Wales) Region or Islands Area (Scotland)
— — — District
—o—o—o— London Borough
· · · · · · · · Civil Parish (England)* Community (Wales)
— — — — Constituency (County, Borough or Burgh)

Coincident boundaries are shown by the first appropriate symbol opposite

*For Ordnance Survey purposes County Boundary is deemed to be the limit of the parish structure whether or not a parish area adjoins

RAILWAYS

Multiple track  } Standard gauge
Single track
Narrow gauge
Siding
Cutting
Embankment
Tunnel
Road over & under
Level crossing, station

TOURIST AND LEISURE INFORMATION

Camp Site — Public Convenience (in rural areas) PC
Caravan Site — Station (special interest)
Parking — Mountain Rescue Post
Golf Course or Links — Youth Hostel
Picnic Site — Information Centre
Viewpoint — Selected places of interest (Manor, SAILING)
Public Telephone T — Motoring Organisation Telephone AA/RAC

*Reproduced with the sanction of the controller of H.M. Stationery Office, Crown Copyright reserved.*

# THE MECHANICS OF MAP READING

**National Grid:** Ordnance Survey maps are covered by a grid or pattern of parallel lines, one kilometre apart. The lines running from north to south are known as eastings and those from east to west as northings. Points on the map can be located by grid references as shown in figure 1. Four-figure references are used to identify grid squares. The first two figures are an easting (vertical line) and the last two a northing (horizontal line). The easting and northing form an L around the square. Six-figure references are used to locate exact points on a map. The first three figures represent the easting and the last three the northing; the third and sixth figures are the number of tenths.

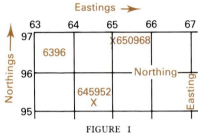

FIGURE 1
**Four and six figure references**

**Scale:** The scale of a map is the relationship between the distance on the map and the equivalent distance on the ground. For example, on a map which has a representative fraction of 1 : 50 000, one centimetre on the map is equivalent to 50 000 centimetres (or ½ kilometre) on the ground.

**Measurement of Distance:** To measure the length of winding roads or rivers, take a straight edge of paper and mark the starting point towards the left end with a sharp pencil; then twist the edge along the course you are measuring, using the pencil at intervals as a pivot.

**Direction and Bearings:** Direction can be expressed (a) by means of compass points, or (b) in degrees as a three-figure bearing measured from north in a clockwise direction. Remember that a compass shows magnetic north, which in Britain is west of true north. The direction of the north–south grid lines is approximately true north, the slight difference being given at the bottom of each Ordnance Survey map.

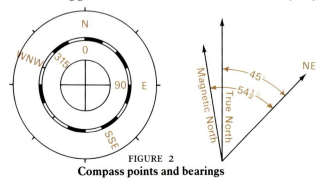

FIGURE 2
**Compass points and bearings**

*Exercise:* Study figure 2, which is incomplete. Make a copy of this diagram and then add all the compass points and bearings which are marked off.

**Setting a Map:** To set a map in the field with a compass, place the compass on the map when it is spread out flat. Then turn the map until the compass needle is lined up with magnetic north marked on the margin of the map. The map is then set. Without a compass, the map can be set if you can locate your own position and a prominent feature both on the map and in the field. A distant church is ideal for this purpose. Turn the map until a line from your position on the map through the identified feature is pointing to the actual feature in the field.

**Contours and Land Forms:** Relief is the shape of the land's surface, and the contour pattern is the main means of interpreting it on a map. Contours are lines joining places of equal height above sea level. The contour interval is the difference in height between adjacent contours. The density of contours indicates the degree of slope. When contours are close together the slope is steep, and when they are far apart the slope is gentle. Concave slopes are steep at the higher part becoming less steep lower down, while convex slopes are gentler higher up becoming steeper lower down. Stepped slopes have a succession of gentle stretches followed by steeper ones. Precipitous slopes are extremely steep, so much so that contour lines may converge and map makers may abandon them in favour of a symbol.

**Cross Sections:** These are scale drawings representing the view as it would appear from the side if a cut had been made through the land surface along a given line. In order to show the relief features clearly, the vertical scale is exaggerated in relation to the horizontal scale. Thus the vertical scale may be 1cm for 100m and the horizontal scale 1cm for 1km. In this case the exaggeration is 1km divided by 100m, which is 10 times.

An accurate cross section involves plotting every contour height along the line of the section, as shown in figure 3. Such a cross section is often used to determine whether one point on a map is visible from another. A sketch section, on the other hand, can be drawn 'by eye', either with a very limited number of fixed points or no measurement at all.

**How to Draw a Cross Section:** Place the edge of a piece of paper along the line of the section on the map. Mark on the paper where the contours cut its edge and number every mark with its height. Then place the paper along the base of a prepared lined paper similar to that illustrated in figure 3. Then erect perpendiculars to the appropriate height and join these up.

**Gradient:** This is the degree of slope. We estimate it by first measuring the distance over which the gradient is required. It may be along a road, in which case remember to measure along the course of the road and not in a straight line. Then work out the difference in height between the two ends using the evidence of the contours and spot heights. If one point is halfway between, say, a 259- and a 274-metre contour, assume the height to be 266·5 metres. If the distance were, for example 2km 300m, with a difference in height of 326m, first bring the distance to metres – 2 300m in this case.

$$\text{Gradient between two points} = \frac{\text{difference in height}}{\text{distance apart}}.$$

The numerator must be reduced to 1, so in this example $326/2300$ is reduced to $1/7·1$. This is the gradient and is usually written as 1 : 7·1 or 1 in 7·1. Road signs now often give

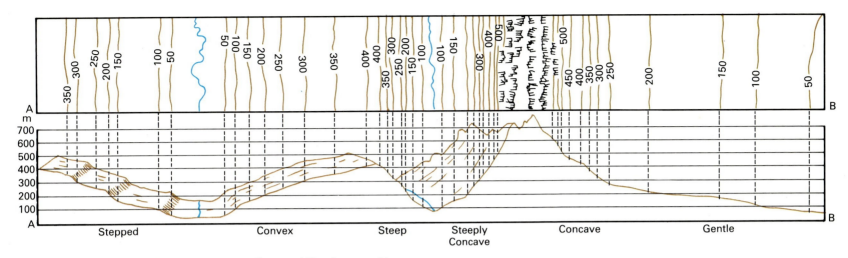

FIGURE 3
**Slopes and sections**

gradients as percentages, so in this example the gradient would be given as 14%.

**Contour Patterns:** When endeavouring to visualise the relief of an area from evidence on the map, use all the information available. This includes not only contour heights, spot heights and triangulation points, but also the drainage and contour patterns. Remember for example that contours are likely to point up towards high land when they bend around a river course.

*Exercise:* Each of the following definitions of land forms is illustrated in figure 4. Redraw this diagram accurately using the help of the grid lines and then label each feature in the correct place. The contour interval is 50 metres; mark the heights against the contours which have not been numbered.

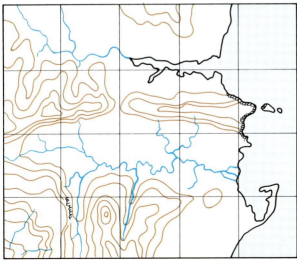

FIGURE 4

**Some Definitions of Land Forms:**
*Ridge.* A long, narrow area of high land.
*Convex slope.* A slope that is steep at the bottom, becoming less steep higher up.
*Concave slope.* A slope that is steep at the top, becoming less steep lower down.
*Promontory.* High land jutting out to sea and sometimes known as a headland.
*Cliff.* A precipitous slope.
*Estuary.* A wide river mouth.
*Bay.* A large coastal inlet.
*Cove.* A small coastal inlet.
*Col* or *Saddle.* A dip between two summits; a high pass through a range of hills or mountains.
*Gap* or *Pass.* A low way through high land.
*Peak.* The highest point of a steep-sided hill.
*Valley.* A narrow depression with a sloping bottom, open at the lower end, and normally occupied by a river.
*Gorge.* A narrow, deep valley with very steep sides, sometimes nearly vertical.
*Delta.* An area of alluvium at the mouth of a river crossed by a number of channels or distributaries of that river.
*Fleet* or *Lagoon.* A shallow stretch of water partly or completely separated from the sea by a narrow strip of land.
*Confluence.* The joining of two watercourses.
*Meander.* The pronounced winding of a river.
*Plateau.* A flat area of high land.
*Knoll.* A low, detached hill.
*Cuesta.* A ridge with a steep slope on one side (known as the scarp slope) and a gentle slope on the other side (the dip slope).
*Spur.* A long, narrow projection of high land into low land.
*Stack.* A small, rocky island near to the coastline and separated from it as a result of marine erosion.

# ROCKS, RELIEF AND DRAINAGE

**The Character of Rocks:** It is generally impossible to determine the actual type of rock in any given area from evidence given on an Ordnance Survey map. We can only make informed suggestions as to what types of rock might be represented. However, with experience it is possible to look for clues which help to identify a probable rock type. Certain types of landscape are associated with particular rocks, and the evidence of mineral workings, drainage patterns and vegetation can also be taken into account.

We sometimes speak of a rock being hard or soft, but of greater significance is its resistance to weathering, which is not the same thing. For example, chemical weathering may be more potent on a hard rock such as an old limestone than on a young rock such as clay which is also very soft. However, mechanical weathering will be more effective on the softer, younger rock. High and rugged ground is mostly built of resistant rocks such as granite, limestone and schists, whilst lowlying land is normally made of less resistant rocks such as clays and marls.

The permeability of a rock reflects the degree to which it allows water to pass through. The absence of surface drainage in this country indicates that the rock is highly permeable; this type of rock may be fissured, as in the case of an ancient limestone, or porous, as in the case of certain sandstones. In the latter case, water can percolate through the minute pores or spaces between the grains of the rock. On the other hand the presence of many streams or of marshland suggests that the rock is impermeable, which means that it does not allow water to pass through freely; slates, granite, and clays are examples.

(a)

**Relief:** In any description of the relief of an area the main aim is to recognise the overall characteristics, rather than the detailed description of individual small features. Consider the map extract on page 59. A brief study should reveal an escarpment cut by the river Witham, with low lying land to the west. A description of the relief might read as follows.

'A prominent west facing, concave scarp slope stretches from the north to the south of the map, broken only by the valley of the river Witham. The land rises to over 60 metres, then falls away more gradually to the east. This gentle dip slope is dissected by a number of west to east trending 'V' shaped valleys (eg. 010682). To the west of the scarp slope there is a low lying vale.'

**Rock Types:**
(a) *Marl:* Keuper marl is a compacted limy clay. The white bands of anhydrite are a relic of dried-up salt lakes. Soils hold moisture and support a good pasture.

(b) *Gravels:* A thick bed of unconsolidated plateau gravels giving infertile soils supporting heather, gorse and bracken with mixed woodland.

(b)

(c) *Chalk:* Soft, permeable rock on which thin, dry soils support a short, tough grass, much of which now is ploughed up for arable farming. (See page 16 for details of Chalk scenery.)

(d) *Sandstone:* A young, loosely consolidated tertiary sandstone, highly permeable and of little agricultural value. Much natural vegetation of coarse grass and woodland remains.

(e) *Carboniferous Limestone:* The thin soils on this permeable, well-jointed rock support a rough pasture for sheep and cattle. (See page 24 for details of Carboniferous Limestone scenery.)

(f) *Shales:* The bedrock is covered by a mantle of mainly shattered rock fragments which provide a poor soil on which a vegetation of bracken and heather develop. The land can be improved to form pasture for cattle and sheep.

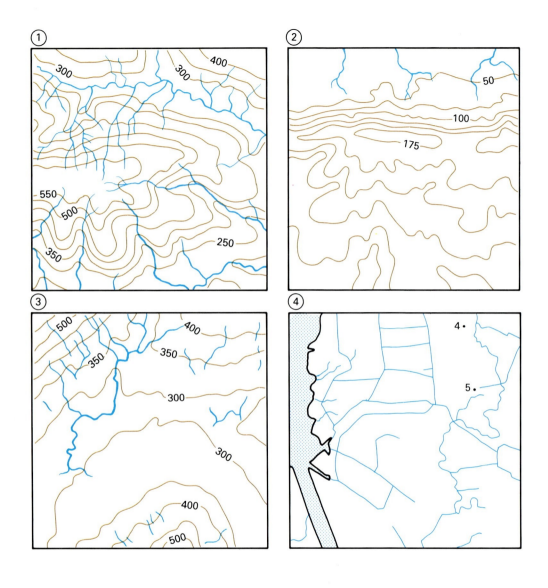

FIGURE 5
**Drainage patterns**

**Drainage:** A description of the natural drainage of an area involves a study of the pattern of streams and rivers. There may be a complete absence of surface drainage as on much of the chalk land. Disappearing streams are a feature of markedly fissured rocks, for example Carboniferous Limestone. Marshland is a result of poorly drained land. Artificial drainage involves either the straightening of existing natural water courses or the digging of drainage ditches. The latter are normally distinguishable from streams by their straightness and right-angled bends.

*Exercise:* Describe the drainage and relief shown on each of the four contour maps. Try to account for the characteristics you note. To help you, key words have been listed below but you will first need to pair these descriptions with the relevant map.

(a) Absence of drainage above 75m; dry valleys; rounded contours; scarp and dip; springs. Chalk downs near Arundel, Sussex.

(b) Discontinuous drainage; swallow holes; plateau; abrupt edges. Carboniferous Limestone of the Pennines near Settle.

(c) Drained plain; straight and winding watercourses; natural and artificial drainage. Fens near King's Lynn.

(d) Dissected upland; plateau top; numerous streams, some incised; main valley in the north of the area. Millstone Grit and shales. High Peak in the South Pennines.

# SETTLEMENT

Settlement reflects the social, economic, and technical changes of the past four thousand years. Relics of late Stone Age settlements, which include hut circles, stone circles, and tumuli, are concentrated on upland areas of southern England particularly the Downs and Salisbury Plain, though much evidence has subsequently been removed by farming.

Settlement became more widespread with the development of a plough suited to the tilling of heavy clay soils. There followed a succession of invasions significantly adding to the size of the population. The Roman occupation provided little lasting effect on rural settlement, but their towns built in strategic locations and linked by efficient roadways have provided the sites for many of our cities of today. The place-name elements of -caster, -cester and -chester are most commonly associated with Roman occupation. The Anglo-Saxons and Scandinavians initially came as raiders and then settled. The Scandinavians restricted themselves to an area north and east of Watling Street. The place-name elements of -ton, -ing, -ington and -ingham are possible indications of Anglo-Saxon settlement; -toft, -by, -garth and -booth are of Scandinavian origin.

**Site:** These early agricultural communities had to be self-sufficient. Their choice of site had to allow access to basic needs which included fresh water, fuel, building materials, arable and pastoral land; there was also a need for a firm foundation for the buildings. The site of a settlement is the spot on which it has grown up. This may be on a river bank or a spring-line if water was the major locational factor. Where security was the principal concern the settlement may have been sited on a gravel terrace or on a defensible hill top.

The site is sometimes indicated in the name of a settlement, as in the example used in figure 6. This small, nucleated village is built on what we call a 'dry-point' site, and this type is commonly found in the fen regions of Norfolk, Lincolnshire, and Cambridgeshire where protection from flooding was the major factor in determining the site. However, evidence on the map indicates that such basic needs as water supply, timber, and farm land were available in the vicinity.

Figure 7 shows the settlement of Fulbeck lying at the foot of a scarp slope, on the banks of The Beck. The village is an example of a 'wet-point' site. *Describe in detail the site of Fulbeck.*

**Position or Situation:** The position of a settlement is its relationship with the surrounding area. It may be at the head of a valley, on a river estuary, or in the centre of a fertile river plain. Initially, communities had to be self-sufficient. Their population was limited by the resources of an area. The limits of the area serving the settlement are represented by the parish boundary. *Can you suggest why the parish boundaries of the two settlements illustrated in figures 6 and 7 are not the same size or shape?* Remember that parishes may have boundaries which take in both lowland pasture and upland rough grazing. The words 'fen', 'common', 'heath' and 'moor' sometimes indicate that the land has been reclaimed and added to the original parish.

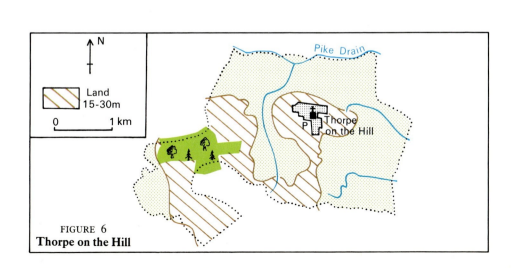

FIGURE 6
**Thorpe on the Hill**

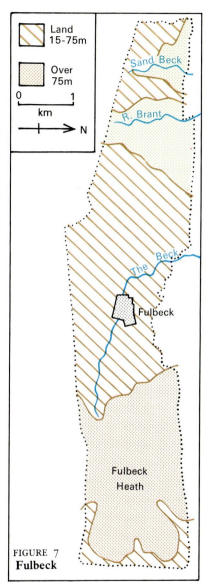

FIGURE 7
**Fulbeck**

*Exercise:* Compare the size, shape, relief and drainage of the parishes of Thorpe on the Hill and Fulbeck.

**Function:** Settlements have a variety of functions or activities. The map of Thorpe on the Hill (figure 6) shows three items of evidence – a church, a post office and buildings. Therefore it is possible to suggest that the village has a service function, a retailing function and a residential function. Most settlements have several functions, some of which may depend on the actual location, e.g. a port function, a mining function, a tourist function. Evidence of 'works' may suggest an industrial function, while a golf course would indicate a recreational function. When you are describing the functions of a settlement, map evidence, including grid references, should always be quoted.

*Exercise:* Study the following list of functions and map features, then pair the correct function with its appropriate piece of map evidence. Function – administrative, tourist, route centre, service, agricultural. Map evidence – principal railway station, hospital, farm, town hall, castle.

**Form:** Form is the shape of the settlement and the layout of the buildings. Many settlements are nucleated, with the buildings closely grouped together. Nucleated villages may have a church, a public house and a water source at the centre, with the oldest residences surrounding them. A more straggling style of village with houses strung out along winding tracks is found in low lying areas, where water was easily obtainable. Dispersed settlement is scattered, such as the isolated sheep farms of the Welsh uplands.

*Exercise:* Study the four maps which show different settlement patterns. Describe each pattern and explain how they differ.

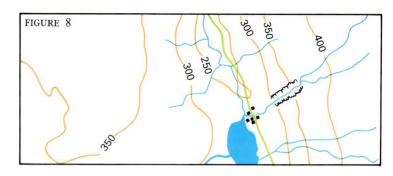

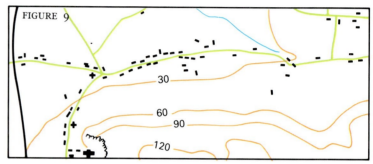

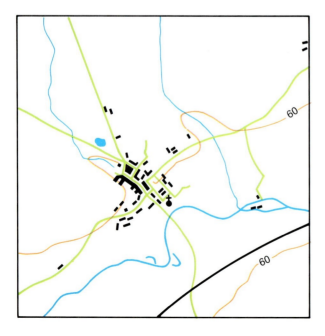

FIGURE 10

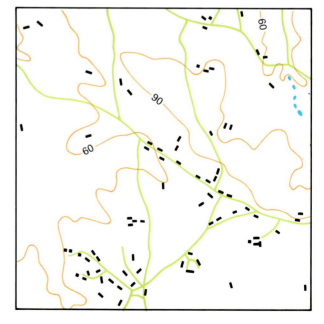

FIGURE 11

# SETTLEMENT PATTERNS

**The Spacing of Settlements:** Just as there is a relationship between the size and function of a settlement, so there is a relationship between the size and the spacing of different orders of settlement. Using an Ordnance Survey map one can recognise patterns of measuring the distances separating hamlets, villages, towns and cities. The table below shows how this has been done for settlement in West Leicestershire.

| Settlement | Average distance apart in km |
|---|---|
| City to town | 20·3 |
| Town to town | 12·0 |
| Town to village | 4·8 |
| Village to village | 3·3 |
| Village to hamlet | 2·9 |
| Hamlet to hamlet | 1·9 |

Why should such a regular pattern emerge?
(1) The relief is level or gently undulating and there are no physical obstacles to settlement.
(2) Villages were once self-sufficient, and if physical and agricultural conditions were uniform they would require the same amount of land on which to grow crops and graze animals. Their spacing would therefore be at regular intervals.
(3) The market towns which grew up during the medieval period had to be within walking distances of the villages. If villages were regularly spaced, then so would be the towns which served them.

**Size and Shape of Settlements' Service Areas:** Where settlements are regularly spaced, the 'tributary areas' of villages and service areas of towns should be of uniform size and shape. If the 'tributary area' were circular this would mean that some land was not taken up (figure 12(a)) or the tributary areas would overlap (figure 12(b)). The most efficient shape that will not leave spaces nor overlap is a hexagon.

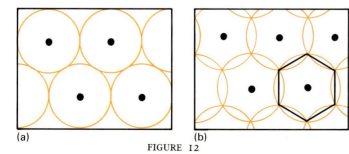

FIGURE 12

Figure 13 shows how a hexagonal pattern can be applied to the service areas of towns and a city. The pattern shows a *hierarchy* of settlements and their service areas.

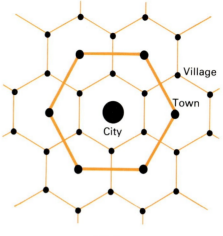

FIGURE 13

In applying this ideal pattern (or model) to the settlements that we can observe on a map, we must appreciate that uniform conditions of relief and agriculture do not always exist. Even so, it is frequently possible to recognise hexagonal patterns as in Leicestershire, as shown in figure 14.

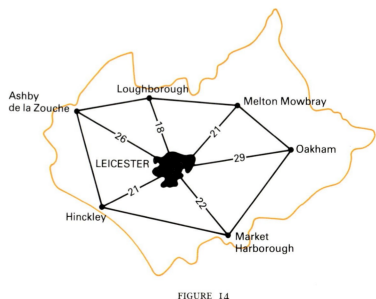

FIGURE 14
**The Leicestershire hexagon**

# COMMUNICATIONS

The term 'communications' can be applied to many types of links which serve people. They may be personal links by conversation or by the postal service. The mass media – television, radio, newspapers, and magazines – are a form of communication. In map reading, although we are primarily concerned with roads and railways, we should not ignore communications by sea, river, canal, air, pipeline, or transmission wires.

**Roads:** The pattern of roads has developed from prehistoric times. The first trackways we have any evidence of followed ridges above ill-drained marsh and thick woodland. Sites of antiquities now often mark the routes of these earliest lines of communication. The influence of the Romans upon our landscape is reflected in their legacy of road networks. Roads such as the Fosse Way and Watling Street, which linked military establishments, took straight paths despite physical obstacles. Considerably improved, these roads remain today as important links in our road network on account of their directness.

Subsequently, roads have tended to avoid areas liable to flooding and areas where steep gradients are met. On a map it is possible to recognise cases where a road has been built on a river terrace above the flood plain, or where it takes a curved path following the contours of the hillside instead of a direct line along the steepest path. This attempt at 'ironing out' the relief is particularly apparent in motorway construction. These roads, designed for fast, long distance journeys have, at considerable cost, effectively used cuttings and embankments for this purpose.

On an Ordnance Survey map, roads are classified according to their condition and function. Motorways by-pass towns which are linked to them by feeder roads. Main roads normally link towns, whereas secondary roads frequently join towns with villages. Other roads have a more local function.

**Railways:** The Ordnance Survey map distinguishes between multiple and single tracks, narrow gauge lines and freight lines, sidings and tramways. Railways are concentrated in industrial areas or corridors linking centres of population. Their routes are to a greater degree controlled by relief than are those for roads because most railway engines are unable to operate over steep gradients. Where railways are unable to follow valleys or coastal plains, cuttings, embankments and viaducts have often been necessary; where cuttings were insufficient to provide a route across an upland, tunnels were dug out.

Figure 15 illustrates how little traffic the railways now carry compared with roads. There may be evidence of this on a map; closed stations and abandoned lines are often marked; these show the attempts to economise by the closure of unprofitable branch lines and little-used stations.

*Exercise:* Comment on the routes of the railway lines shown on the Medway extract.

**Rivers:** It is not always possible to discover the extent to which rivers serve as lines of communication. Certainly many were once far more important than they are today. However, with the size of ships increasing and the more efficient alternatives to river transport being developed, rivers and riverside docks have tended often to fall into disuse. Reference to wharves, locks and other installations may provide evidence that the river is or has been used by certain types of craft.

**Canals:** Although some canals are still used by barges carrying bulky goods, most are used only by pleasure craft today. Their construction during the early stages of the Industrial Revolution was made possible only by the use of aqueducts, embankments, cuttings and tunnels to obtain a level course, and locks to enable movement from one level to another.

**Air Communications:** Although air routes are not shown, airports which act as passenger terminals are named and airfields which are used by the Air Force or clubs are similarly marked as such. Runways for aircraft are indicated by broken lines. They need to be sited on flat ground and, because of the space they require and the noise produced, they are usually sited some distance from built-up areas.

**Description of a Route:** In a description of a route, reference should be made to distance, direction, elevation, slope and the nature of the landforms crossed. You should discuss the ways in which the route negotiates natural obstacles such as steep gradients, flood plains and watercourses by such means as tunnels, cuttings, bridges and embankments.

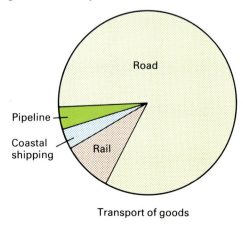

Transport of goods

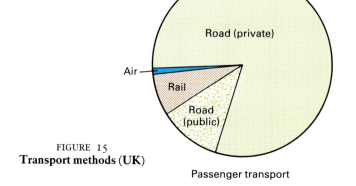

FIGURE 15
**Transport methods (UK)**

Passenger transport

# ACCESSIBILITY AND CONNECTIVITY

**The Accessibility of Places in a Network:** The efficiency of communications depends upon the ease of access to places within a network.

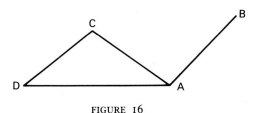

FIGURE 16

In the above figure which shows a network of communications, A is the most accessible point because you can reach B, C or D by means of a single journey. To reach D and C from B involves two journeys. We use the word 'link' to describe the journey between two places. The number of links needed to reach the furthest point in the network by the shortest route is called the accessibility number.

**Example:** To discover which is more accessible, Hinckley or Leicester.

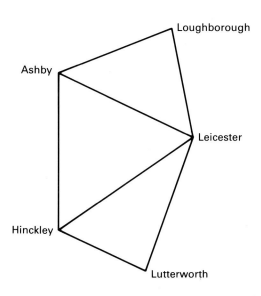

FIGURE 17

Calculations

| Accessibility number for Hinckley | Links |
|---|---|
| Hinckley to Loughborough | 2 |
| to Ashby | 1 |
| to Leicester | 1 |
| to Lutterworth | 1 |

Therefore the accessibility number for Hinckley is 2.

| Accessibility number for Leicester | Links |
|---|---|
| Leicester to Loughborough | 1 |
| to Ashby | 1 |
| to Hinckley | 1 |
| to Lutterworth | 1 |

Therefore the accessibility number for Leicester is 1.

Therefore Leicester is more accessible than Hinckley.

*Exercise:* Calculate the accessibility of Ashby and Loughborough in this network.

The method we have used is a very simple one for assessing which is the most or least accessible place in a network. Other techniques can be used such as:
(a) adding up the number of links from one place to all the other places in the network eg. for Leicester the value is 4.
(b) substituting in method (a) the actual distances.

**The Connectivity of a Network:** Connectivity means the number of links between places in a network. The more links there are, the greater the level of connectivity. A simple way of determining the level of connectivity is to divide the number of links by the number of places.

$$\text{Connectivity Index} = \frac{\text{Number of links}}{\text{Number of places}}$$

Connectivity index values can then be compared with the table below.

| Index | Characteristics | Example |
|---|---|---|
| Less than 1 | Very simple, no circuits | X |
| 1 | Connected, one circuit | Y |
| More than 1 | More complicated networks | Z |

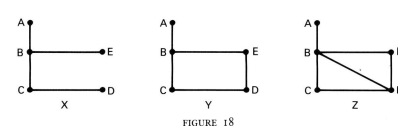

FIGURE 18

## QUESTIONS ON THE MEDWAY MAP

### Exercise A
1. (a) Name the long distance path at 726640.
   (b) Why are there few surface streams shown on the map?
   (c) What type of relief feature is in square 7664?
2. Give six figure references to locate the beacons in the mouth of the estuary of the Medway. Why are they there?
3. Draw a plan of the section of the motorway which appears on the map. Mark on this plan all the features associated with the motorway (e.g. sliproads, cuttings, bridges, viaducts). Label these features.
4. What problems were encountered by the constructors of the M2 judging from the evidence on your plan? Refer to page 12 for assistance.
5. Explain the presence of an airport at 744645. Again, refer back to page 12.

### Exercise B
1. Describe the character of the estuary downstream from the pier at 755682.
2. Construct an annotated section from Gibraltar Farm (779633) to 800690. Mark on your section built-up areas, main and secondary roads, woodlands and orchards.
3. Comment on the land use along the line of your section. Is there any correlation between relief and land use?
4. Describe the location of the built-up areas and then make a list of the factors which account for this location.

### Essay Work
On a map of the British Isles mark the course of all the motorways. Label each of them.

What effect have motorways had (a) in easing congestion on our roads and (b) in giving access to previously remote areas?

### Decision making
A major hotel chain has decided to build a motel in the Medway area. As an executive of this company it is your task to select the location for the new motel. Five available sites have been short listed as being of a suitable size and layout. The location of each site can be found from the following map references.

Site A—715654  Site B—717695  Site C—747650
Site D—795687  Site E—795665

(a) What factors must you consider when locating a motel?
(b) Which of the five sites that are available would be most suitable? Explain your choice.

*Photograph Aerofilms Ltd*

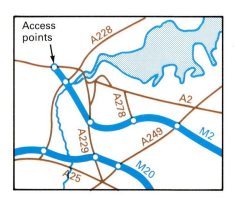

FIGURE 19

**Road network of the Medway area**

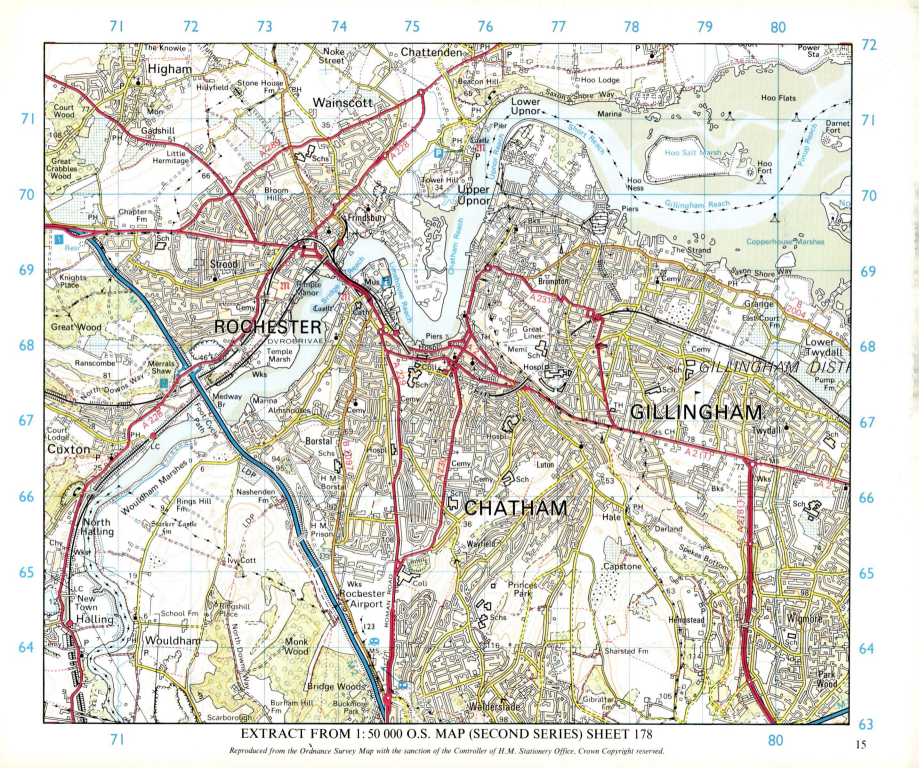

EXTRACT FROM 1:50 000 O.S. MAP (SECOND SERIES) SHEET 178

Reproduced from the Ordnance Survey Map with the sanction of the Controller of H.M. Stationery Office, Crown Copyright reserved.

# SHOREHAM AND THE SOUTH DOWNS

**Chalk Scenery:** Chalk is one of the most easily recognised rock types. It is a soft, white limestone and has minute pores and thread-like fractures or cracks so that water passes through very slowly. For example, a boring into chalk may take many days to fill with water although the borehole goes below the water table into water-saturated rock. Nevertheless, the rock is sufficiently permeable that chalk uplands generally lack surface drainage. However, rivers do cut across chalk uplands maintaining a course which was taken long before the present relief pattern evolved; they do so in deep valleys which form gaps through the chalk uplands.

Most of the British chalk uplands take the form of cuestas; these have a steep slope, known as the scarp or escarpment, and a gentle or dip slope. The overall relief outline is one of smooth, rounded curves with convex slopes dominating. The dry valleys are a feature of chalk country; those that cut into the scarp slope are short and steep sided with an abrupt upper end, and are called coombes in some parts of the country. On the dip slope, the dry valleys are longer and shallower with tributary valleys leading off. Temporary streams, known as bournes or winterbournes, may flow in the 'dry' valleys for some months of the year when the water table is high. The water table, which is the upper level of saturation of a permeable rock, will be higher after a wet spell of weather. At the foot of both the scarp and the dip can be traced a spring line; this is normally at the junction of the chalk with underlying impermeable rocks.

Soil that develops on chalk is thin and is covered with a short, tough grass with occasional areas where box and yew grow. Also on some chalk areas there are clumps of beech; these have shallow roots which radiate out for considerable distances. Until recent years, chalk lands were used almost exclusively for pasture, particularly for sheep. Where surface water is limited, small round ponds known as dew ponds are numerous; some of these are natural but many are lined with concrete. These watering points for livestock are also numerous in other limestone areas. Today, the gentler slopes are regularly cropped, although both cattle and sheep are grazed on the leys. Farms are large, extending often to over 500 hectares, and barley is the chief crop. The soil is naturally of low fertility so that considerable use is made of fertilisers.

Chalk uplands are very thinly settled, villages developing on the spring line. Larger villages and towns tend to develop at gaps in the chalk.

**Clay Scenery:** Clay is a finely grained, soft rock which is easily eroded and holds moisture. It does not have joints or bedding planes and once it is saturated it is impervious. Claylands are normally undulating lowlands with much surface drainage. Rivers meander and are likely to build up alluvial coatings near their banks due to periodic flooding; here the relief will be very flat, the borders often being marked by river terraces.

The claylands were once either covered with thick, deciduous woodland, oak and elm being the dominant species, or, where drainage was poor, they were waterlogged marshy areas. Over the centuries these lands have been drained and cleared, and now, although hedgerows give a well-wooded appearance to the countryside, mere copses are the remaining relics. The heavy soils which develop on the clay are more expensive and difficult to prepare for arable crops than light soils, but they do support a rich grass. Consequently these areas tend to be mixed farming regions, individual farmers

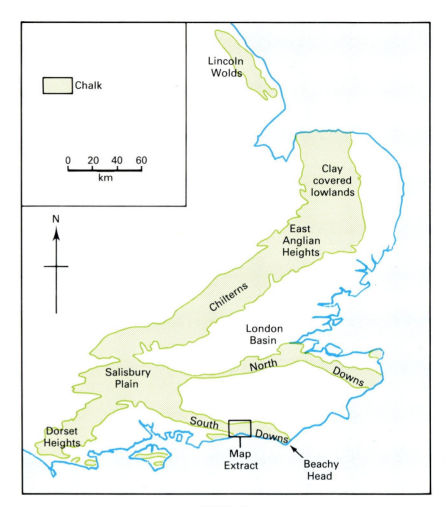

FIGURE 20
**Chalklands of South and East England**

specialising perhaps on beef production, dairying, sheep farming or, less often, on crops. Settlements avoid the lower-lying areas, but frequent village sites are slight elevations such as gravel pockets and terraces.

**The Shoreham Area:** The map extract over the page is of a portion of the South Downs in Sussex together with a coastal plain immediately to the west of Brighton. *First read the general account on chalk scenery and then consider the following clues which may indicate that the area on a map is chalk. Individually, these clues are certainly not necessarily conclusive.*

Cement works (also in other limestone areas), chalk pits, place names ending in coombe or bourne, white horses cut in the rock, many references to tumuli and barrows (but these are found on many other open uplands), the terms 'down' in south-east England and 'wold' in Yorkshire and Lincolnshire, normally a complete absence of surface drainage but many dry valleys, dew ponds.

*Next, study the map extract and identify the upland area and then find eight clues which, taken together, prove that the uplands are made of chalk.*

North of the Chalk scarp lies the Vale of Sussex. Only a small portion of this appears on the extract, but there is sufficient for you to identify an area of undulating lowland which has been produced by the wearing down, mainly by running water, of the relatively soft, impermeable clay. Nearer the River Adur, in this northern portion of the map extract, the scenery is different; the land is much flatter, artificial drainage is needed, and there is a considerable stretch where there is an absence of settlement. *Read again the general account on clay scenery and then explain why there is this difference between the undulating land and the flatter tract in the Vale of Sussex.*

South of the South Downs lies a fertile coastal plain formed on alluvial and other recent deposits. It was formerly noted for market gardening including the hothouse cultivation of vegetables and flowers. Is there any evidence on the extract that this activity still exists? The spread of housing has converted a number of separate towns, including Brighton, Shoreham, and Worthing, into a sprawling coastal conurbation. In the main this is a residential and holiday area. *Study the map extract and identify seven different symbols that represent features a tourist might use or be interested in visiting.*

*Shoreham harbour* lies behind a shingle spit which grew eastwards across the estuary of the River Adur as a result of the west to east longshore drift. Sand and mud accumulated on the landward side, and in the course of time salt marshes were formed. In 1760 the entrance to the port was at Aldrington at the eastern end of the present harbour. Shortly afterwards a cut was made in the shingle spit farther to the west. A second cut was made in 1821 on the site of the present harbour entrance.

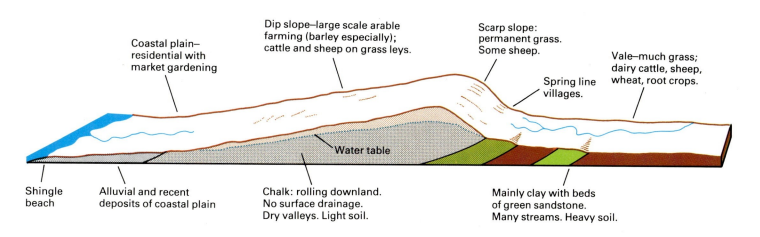

FIGURE 21
**Section across the South Downs**

## QUESTIONS ON THE SHOREHAM MAP

### Exercise A
1. Which of the following land forms – plain, dry valley, scarp, gap, spur – is to be found at 262075, 265112, 200055, 235112, 200080? See page 5 for definitions of the terms used.
2. Quote, giving grid references, TWO pieces of evidence for prehistoric life.
3. You meet a motorist at Fulking (248114) who asks you the way to Bramber Castle (185107). He has no map or compass. Draw a simple sketch map of the route he requires, containing the necessary directions.
4. Describe the view from 212074 looking west and comment on the routes followed by the roads.

### Photograph question
After you have studied the photograph carefully make two lists. On the first list include the information which a photograph shows but which a map cannot; on the second list give the information that the map shows but which a photograph cannot.

### Exercise B
1. Draw a map of the area covered by the Shoreham extract but on a scale of 1:100 000 to show a division of the area into three geographical regions. Name the three regions, insert the 76m contour, the River Adur, and print the words SCARP SLOPE and DIP SLOPE in the correct places.
2. Describe the distribution and type of settlement in each of the three regions you have marked. Study the section on the previous page and then explain how the physical geography has influenced the settlement.
3. Describe the physical features of the River Adur and its valley from Bramber to the sea.
4. Comment on the presence of (a) a castle at 185107, (b) a golf course in 2610 and (c) locks at 242048.

*Photograph Aerofilms Ltd*

### Essay Work
Draw a map to show the chalk outcrops of England and, with the aid of your atlas, name as many of the hills formed of chalk as possible. Also name any clay vales that you can find.

What are the main differences between chalk and clay in respect of the nature of the rock and the physical and human features of the landscapes?

### Interpreting a section
Use the map to identify the features labelled A, B, C, D, E, F and G on the section (figure 22) which is drawn from 260130 to 260045.

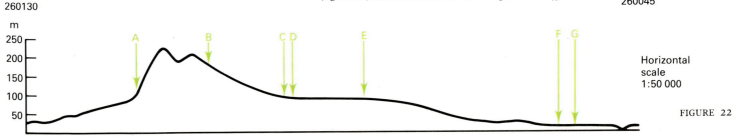

FIGURE 22

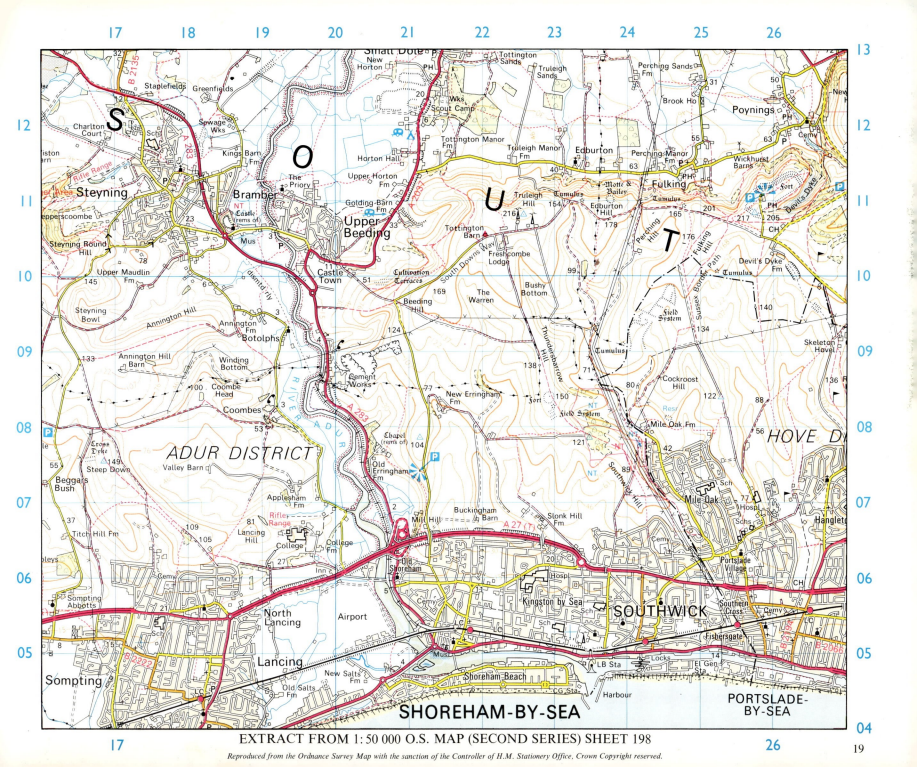

# HULL

**The Development of Hull as a Port:** There was a port at the mouth of the River Hull in 1160. In 1280 it was acquired by Edward I, renamed Kingston-upon-Hull, and it soon became the third port of the kingdom. It remained a river port based upon the estuary of the Hull river until 1809, when the Humber Dock which opened directly onto the Humber came into use.

The Hull estuary had provided a sheltered site with ample quay space for the small craft of the Middle Ages. On the other hand, the land here was subject to flooding and much of the surrounding area was marsh. With the early stages of the Industrial Revolution midway through the eighteenth century, there was a much greater demand on the port of Hull for here was a point where the larger ocean-going vessels could transfer goods to the narrow canal boats for transport up the Aire–Calder systems to the new industrial regions of the West Riding of Yorkshire.

The first docks to be dug out from the soft clays and alluvium followed the line of the old outer wall and moat which had protected the old city; these were Queen's, Princes, and Humber Docks. By 1850 much larger docks were needed to cope with the trade created by the nearby industrial regions of the West Riding and Midlands, and great expanses of the unoccupied marshy land to the east and west of the old city were taken. To the east, Victoria Dock was constructed, while to the west a long line of docks was formed by excavating enormous trenches, the material removed being used to build up a secure embankment between the docks and the Humber estuary. The more recent expansion has continued downstream of the city centre. The King George Dock with its extension, the Queen Elizabeth Dock, has three kilometres of quayside including new and extensive facilities for container and vehicle traffic. Over half a million passengers now pass through the port each year. The extensions downstream have been at the expense of the older docks and many of these are now closed. The rapid decline in the fishing industry also contributed towards their disuse.

*Exercise:* Read this short account on the development of Hull as a port and study the maps. Then explain the advantages and disadvantages which have influenced Hull's development. Consider both features of site (there are at least three) and also position in relation to hinterland and direction of overseas trade.

**Hull as a Manufacturing Centre:** The main manufacturing industries of Hull developed where imported raw materials could be readily obtained from the docks. Therefore the sites of these manufacturing concerns have concentrated alongside the docks in a belt some ten kilometres long while another belt extends on either side of the River Hull for some four kilometres. The finished products of these factories include margarine, soap, glue, rice, fish meal and fertilisers, flour, paint, and chipboard. *What raw materials are required for these factories and where would these products be imported from?*

There are new industrial pockets on the outer fringes of the city. To the east the Salt End industrial estate uses the oil and other liquids piped ashore into the depots there for the manufacture of chemicals, plastics, and textile fibres. Other new industrial concerns also provide alternatives to the earlier concentration on the processing of

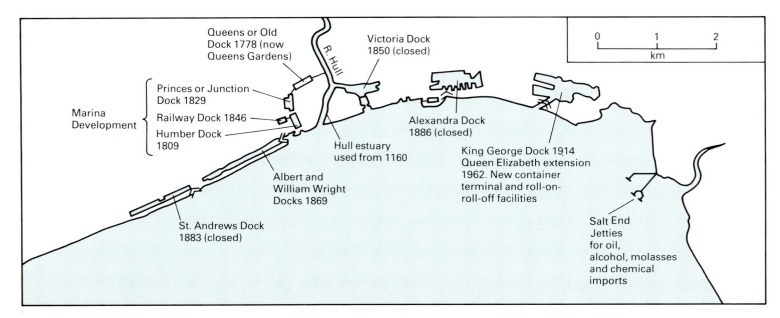

FIGURE 23
**Hull Docks**

imported vegetable products to give a broader manufacturing base; they are concerned with light engineering and clothing products.

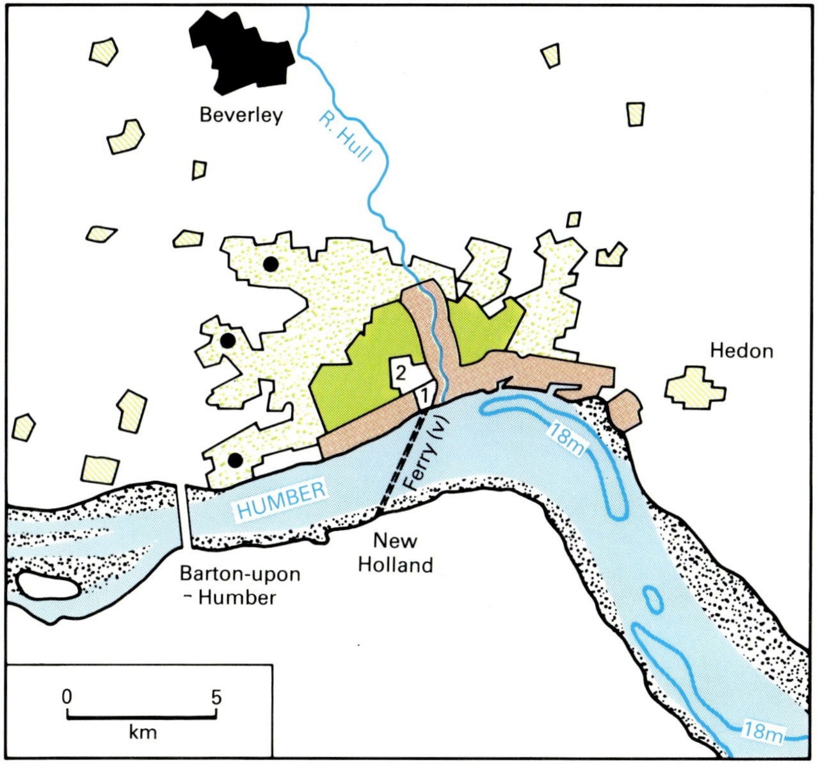

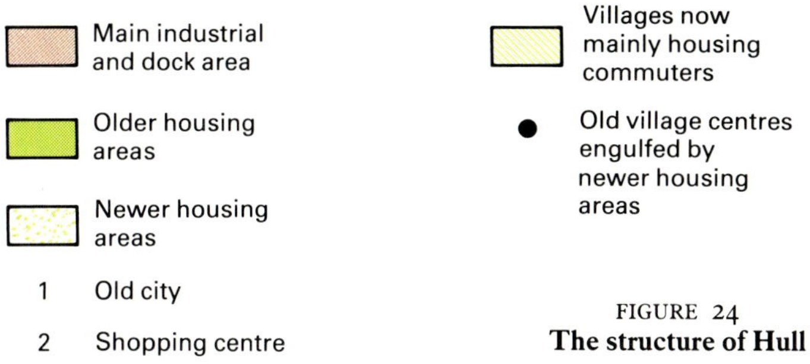

*Exercise:* Study the six patterns illustrated in figure 25 and then attempt to identify these from the extract. Suggest other patterns on the extract which are similar and give the references. The simplified structure plan of Hull should help.

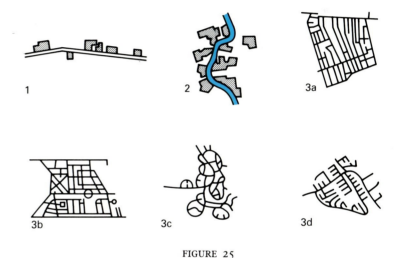

FIGURE 24
**The structure of Hull**

FIGURE 25
**Settlement patterns in Hull**

(1) Ribbon-like industrial development which can use the services of the adjoining main road together with the dock facilities the other side of the road.

(2) Similar large blocks but this time served by railways and extending alongside the River Hull which provides barge quayage on the plant's site.

(3) Various residential patterns. In these three diagrams only the road pattern is indicated.

(a) Many long lines of streets; rectangular, gridiron pattern; near to the city centre and the older docks. This suggests the long-established housing areas.

(b) Geometrical patterns, very ordered and balanced, consisting of squares, diagonals, and circles; located on or near the edge of the city. These suggest integrated, planned housing areas; the unimaginative patterns are typical of many housing estates built in the period 1930 to 1950.

(c) Flowing, rounded, irregular road patterns with crescents and cul-de-sacs forming compact residential areas of the type found on the edge of many large towns; generally constructed since 1950.

(d) Regular road patterns with many cul-de-sacs, located on the edge of the town. Undisturbed closes and an emphasis on increased density are typical of housing built since 1970.

**The Settlement Patterns of Hull:** Normally, one cannot positively identify types of settlement from an Ordnance Survey extract, but the evidence available can give pointers to the types that may have developed.

# QUESTIONS ON THE KINGSTON-UPON-HULL MAP

**Exercise A**
1. (a) Give six figure references for the main railway station and the Hull terminus of the vehicle ferry that crosses the River Humber.
   (b) Identify the symbols at 164268 and 120310.
2. (a) Identify four types of road in square 1432 and then illustrate and label each of them.
   (b) Identify two types of railway in square 0629 and then illustrate and label each of them.
3. (a) Calculate the width of the River Humber from Salt End Jetties to Skitter Ness.
   (b) State one disadvantage of this distance for Kingston-upon-Hull.
4. Compare the watercourse which runs south from 083340 with that running south from 127340. In what ways do they differ?

**Practical Work**
Study the photographs, then draw a plan of the area covered by the top photograph. Mark on the docks, jetties, industrial areas, agricultural land and indicate the passenger berth and container terminal.

**Exercise B**
1. Study the settlement patterns in the vicinities of 070330, 080280, and 118293. One is an industrial block, one an old housing area, and one a new housing estate. Pair the descriptions with the grid references and give reasons why you do so.
2. Study the industrial development in square 1627. Then, using the evidence on the map only, describe the complex and explain why it might be located there.
3. Describe the functions of Kingston-upon-Hull using evidence on the map only. Refer to page 10 for guidance.
4. Account for the location of the mud flats exposed at low tide.

**Essay Work**
In 1972, 2420 trawlers were based at the port of Kingston-upon-Hull. By 1982, this number had fallen to 224. Account for this dramatic decline.

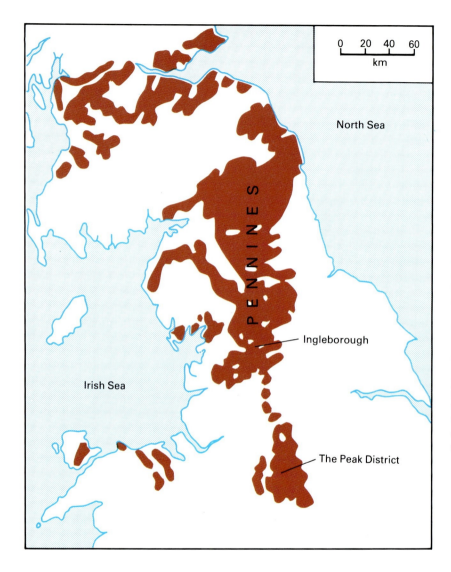

FIGURE 26
**Carboniferous Limestone areas of Northern England**

# INGLEBOROUGH

**Carboniferous Limestone Scenery:** The Carboniferous Limestone series in Great Britain is over 1 000 metres thick but not all of this is limestone. In the series, beds of limestone are interspersed with layers of less resistant shales and sandstones. The limestone itself has well-developed systems of joints and bedding planes and is consequently permeable. It is very resistant to mechanical weathering but chemical weathering plays a significant part in the development of limestone scenery. Rainwater absorbs carbon dioxide from the air to become a very dilute carbonic acid which reacts on the limestone, carrying away small quantities of calcium carbonate in solution. This process has an obvious effect where limestone beds are thick; here water follows joints and bedding planes in the rock, enlarging them and creating a labyrinth of underground watercourses, many of which have been subsequently abandoned.

The limestone uplands provide a rolling, plateau-like landscape. The upper surface is devoid of drainage but rivers occupying deep troughs have cut the uplands into blocks. Even shallow dry valleys frequently are bordered by precipitous edges of limestone. The open, light appearance of the uplands is due partly to the absence of trees and partly to the light grey colour of the weathered rock where it outcrops as scars or pavements. In the areas used for farming, this appearance is heightened by the dry stone walling surrounding the many small fields.

The thin soils of the uplands support a coarse grass which provides adequate pasture for the hardy, upland breeds of sheep. In contrast, the steep valley sides are sometimes thickly wooded. Villages are normally located in the valleys, but the plateau surface over much of the limestone country is dotted by isolated farms often sited in shallow hollows. Where the valleys are deep and winding, roads tend to avoid them, keeping rather to the plateau, but where the rivers have opened up wide vales a typical communication pattern of twin roads, each close to one side of the vale, has emerged. *Why do you think this is so? Consider winding rivers, cost of bridge construction, water meadows, and the location of farms built on dry sheltered sites close to springs.*

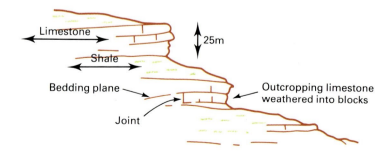

FIGURE 27
**Field sketch of an outcrop of Yoredales**

**The Ingleborough Area:** The Ingleborough extract illustrates many of the features of Carboniferous Limestone. The crest of Ingleborough is capped with a layer of Millstone Grit, but most of the steep slopes immediately below this capping are made up of a series of rocks called Yoredales (see figure 27). These consist of repeated bands of limestone, shales and sandstone which give a stepped appearance, the limestone outcrop forming the scar and the less resistant shales and sandstones forming the bench as is illustrated in the field sketch. This type of outcrop is typical of much of the limestone scenery throughout the Pennines.

Below the Yoredale slope there is a great shelf of massive limestone which forms a plateau just above 400 metres high. This particular layer of rock is known as Great Scar Limestone and its outer edge forms precipitous scars which provide vertical or near vertical drops of over 150 metres in places. The joints are very marked and the layer is uninterrupted by other impermeable beds. Consequently, streams running down from the upper rocks, when they reach this shelf, soon plunge into swallow holes some of which are vertical drops of up to 120 metres. The streams emerge at the surface lower down just above the impermeable, older rocks on which the limestone rests. These older rocks have been exposed by erosion in the valley floor of the River Greta.

The Pennines were covered by ice sheets during the Ice Age, and although there are not such obvious features of glaciation here as in the mountainous areas of the country, nevertheless many of the major Pennine vales, including that of the Greta, have been deepened by ice action.

*Exercise:* Now that you have read the general account on limestone scenery and the specific account on the Ingleborough area, study the map extract of the area. Then consider the following clues which may indicate that portions of the extract are made of Carboniferous Limestone. List those which apply, giving grid references.

Disappearing streams; collapsed caverns; gorges; the use of terms such as swallets, swallow holes, potholes, and scars; limestone quarries and lime works (also present in other limestone areas); dew ponds (particularly numerous in farmed areas of the Carboniferous Limestone).

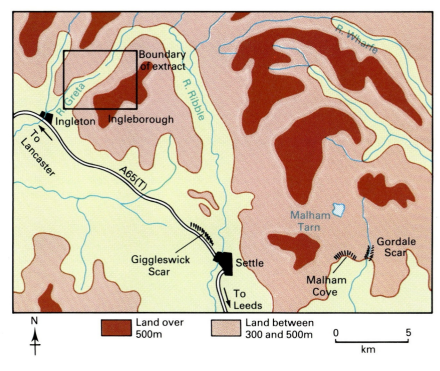

FIGURE 28
**The Central Pennines**

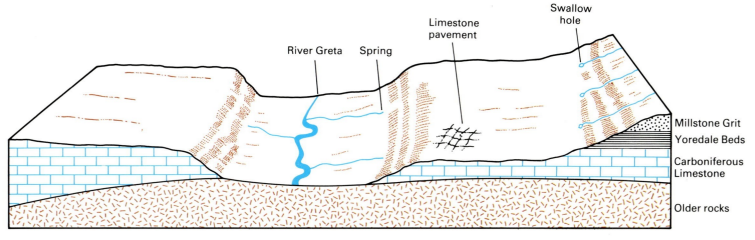

FIGURE 29
**Section across the Greta Valley**

## QUESTIONS ON THE INGLEBOROUGH MAP

**Exercise A**
1. Describe the land surface at (a) 735755, (b) 730771, (c) 744764.
2. Give a detailed description of the River Greta and its valley. The following suggestions may help. Is the valley wide or narrow? How wide? At what point? What is the relief of the valley floor and how steep are its sides (for example there may be a gradient equivalent to an ascent of 300m in 1km)? Is the river fast flowing? Does it wind or meander? Are the tributaries lengthy? Which way does the river flow? How does the valley differ in square 7377 from 7175?
3. Describe the courses of the two roads in relation to the relief and drainage courses.
4. Name three different human activities practised in the area, in each case quoting map evidence and giving grid references.

**Practical Work**
On a grid showing all the eastings and northings and on the same scale as the extract, mark in the 900-foot and 1 800-foot contours and all the drainage systems shown. Between what levels do you find surface drainage and between what levels is there no evidence of surface drainage? What can you deduce from this?

**Exercise B**
1. Draw an accurate section along the line from 710780 to 750740. Mark on it the River Greta, Ingleborough Hill, and any scars crossed.
2. Along the line of the section calculate the degree of slope at certain stages to illustrate changes in the relief. For example, you might take (a) the slope on Great Hard Rigg Moss between the footpath and the top of the scar, (b) the slope from the top of the scars to the road near Sprincote, and (c) the slope from 738750 to the platform edge of Ingleborough.
3. Describe the general features of the relief of the map using the evidence of your section and calculations of slopes.
4. In what types of locations do you find the following features: scars, pot holes (of the type mentioned on the map extract), caves, springs and mosses?

**Essay Work**
Write an account describing and explaining the distinctive characteristics of Carboniferous Limestone scenery. Use examples from the Ingleborough area if you wish and draw diagrams to illustrate your work.

**Photograph Question**
Identify and describe the scenes in the two photographs which were both taken from Keld Bank (747773).

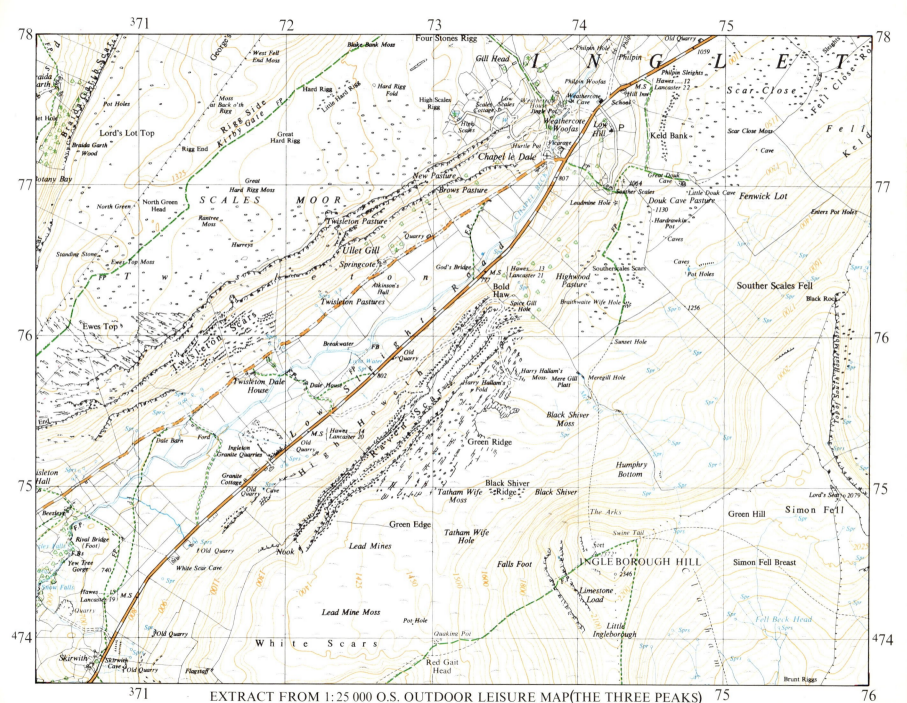

EXTRACT FROM 1:25 000 O.S. OUTDOOR LEISURE MAP (THE THREE PEAKS)

# DOWNHAM MARKET AND THE FENS

**Drainage Schemes in the Fens:** The Fenland was formerly a vast marsh in which grew sedges, reeds and similar plants. The decay of this vegetation led to the growth of a thick layer of peat. Near the Wash, the peat has been covered subsequently by marine silts. Reclamation of this marshland has been progressing for many centuries. Early schemes were small scale and sporadic; they generally extended out from the winding ridges of land which cross the fens and which provided many of the early sites for settlement. Today it is possible to distinguish on a map between the long-established, smaller, irregular drainage patterns and the more recently made long, straight cuts and drains of much larger schemes. The road patterns of the two types, that is the long-established and the more recent, are very similar to the drainage patterns.

One of the earliest large schemes was undertaken by the Dukes of Bedford in the seventeenth century when a large tract known as the Bedford Level was reclaimed. The meandering River Ouse was partly replaced by straight cuts known as the Old Bedford River and the Hundred Foot River in order to carry water more quickly to the sea. The land between these cuts was left as a wash, that is an area which could be flooded so as to hold back water and prevent flooding farther downstream.

A great step forward came with the introduction of steam pumps in place of the old wind pumps. This was in the early nineteenth century. Many of the fields are below river level and pumps are needed to lift the water from the drainage canals into the river. Today the stream pumps have been replaced by diesel and electric pumps.

The drainage systems which appear on the map extract of the Downham Market area merit special attention. In 1953 the River Ouse overflowed, breaking its banks at places between Denver and the sea. In 1956 a scheme was put into operation in which a straight channel immediately above Denver sluice was excavated; this was dug alongside the Great Ouse, joining up again with the main river very near its estuary and just above King's Lynn. The function of this channel is to ensure that flood water does not build up at the crucial Denver junction of watercourses. The Ouse, for several kilometres above Denver, has been widened to improve the flow of water from the Ely area to the Wash. Also, in 1960 a 'cut-off' channel was excavated in order to carry excess water from the upper courses of the Lark, Little Ouse, and Wissey to the new relief channel at Denver. In 1971, the Ely–Ouse–Essex scheme was completed. By this scheme, water from the Great Ouse is directed southwards down the cut-off channel and then by pipeline south-eastwards to Essex where water demand has overtaken supply.

**Farming and Settlement in the Fens:** Now the Fenland is one of the most productive farming areas in Britain. The deep, fertile soils have encouraged farmers to concentrate on intensive forms of arable farming, and about 80 per cent of the land surface is used in this way. The little permanent grassland that exists is mainly found on the washes and along the river banks. There are very few cattle and sheep. Although the farms are small, about half of them being below 25 hectares, the farmers' incomes are above the average for the country as a whole. This is because they concentrate on the production of valuable cash crops and use intensive methods of farming. Cash crops are those which are grown for sale as opposed to those which are used as fodder for livestock kept on the farm. The intensive methods used include heavy manuring and

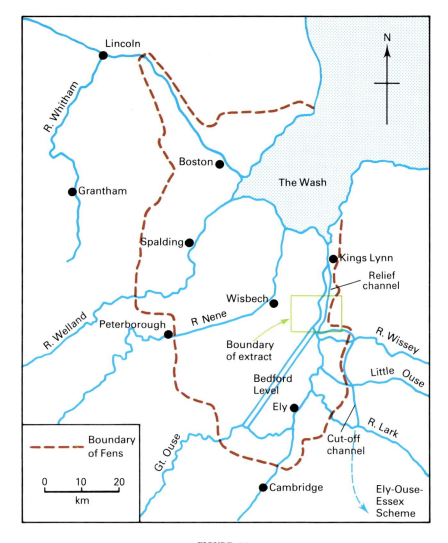

FIGURE 30
**The Fens**

careful cultivation which give a high yield per hectare. Wheat, sugar beet, and potatoes are the most important general farm crops, but throughout the fens there are scattered but very important pockets of market gardens and orchards.

Most of the older towns and villages are located on slightly rising ground at the edge of the Fenland or on 'islands' where gravel or clay pockets form low mounds or ridges which were built up by the old courses of rivers. However, much of the recent settlement is dispersed and straggles along roads so that often one village merges into the next.

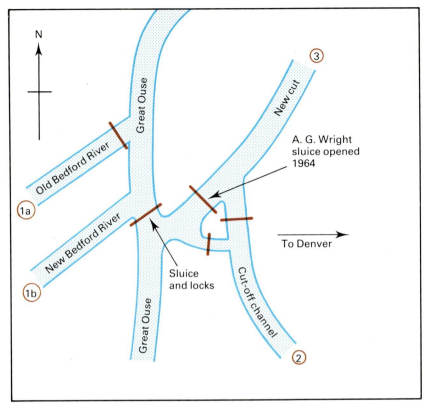

**FIGURE 31**
**The Denver Sluice complex**
What is the function of 1a, 1b, 2 and 3?

**Land Use Transects:** In order to make a quick and simple appreciation of the land use of an area, a land use transect showing crops grown along a set line may be of value. It might only be possible to do this along a roadside, but such a transect may give a distorted impression. In the Fens, for example, one would expect to find a greater proportion of the most intensively cultivated crops, including market garden and greenhouse production, close to the farmhouse, which could be one of many alongside the road. The transect should be at least two kilometres long and one can pace the length for each crop or merely count the land use of each field. However, in the Fens this latter method is useless since the land is frequently cultivated in strips which may be only 20 metres wide and perhaps 500 metres long.

*(1) Notes made on an eight-kilometre transect on land to the north-west of Barroway Drove on the Stow Bardolph Fen.*
In this area the community lives in a linear village strung out along the Barroway Drove for a distance of over 4 kilometres. The community has its own junior school, garage, public house, chapel, and recreation ground. The land on Stow Bardolph Fen has no marked speciality as is found, for example, around Wisbech, where there are considerable pockets of land occupied by market gardens and orchards, or Spalding, where market gardening and bulb and greenhouse production cover considerable areas. Here on Stow Bardolph Fen general farm crops are intermingled with market garden crops and every available space is used; for example, strips of wheat no more than 15 metres wide were being grown between houses on the roadside.

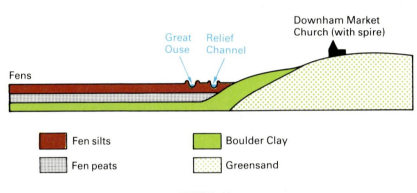

**FIGURE 32**
**Section across the area covered by the Downham Market extract**

Land use along the transect given as a percentage: wheat 27; barley 14; potatoes 20; sugar beet 15; onions 3; cabbages 3; lettuce 2; beans 3; beetroot 1; strawberries 6; sweet peas 1; greenhouses and their surrounds 3; pasture and waste 1; playing field 1.

*(2) Notes made on an eight-kilometre transect on land to the east of Downham Market, on the Greensand outcrop.*
This area provided a very different scenery to that on the fen silts. Here we are back to tall hedgerows and mixed farming. The less fertile Greensand has considerable remnants of deciduous woodland, and there are plantations both of deciduous and coniferous trees. Settlement is concentrated in a number of nucleated villages but there are a few large scattered farmhouses.

Land use along the transect given as a percentage: wheat 24; barley 8; pasture 48; woodland 20.

*Exercise:* Make two pie graphs to illustrate the two sets of statistics. Start by drawing two circles each with a radius of 4cm, multiply each of the given percentages by 3·6 and then divide the circles into slices using the figures you have calculated. Using a protractor, each figure will be used as an angle from the centre of the circle. Shade each segment in distinctive colours, using the same colour scheme on each diagram, and label.

*Photographs Aerofilms Ltd*

## QUESTIONS ON THE DOWNHAM MARKET MAP

**Exercise A**
1. What is the general height of the land west of easting 60? To do this work out the average height of all the spot heights you can find. What is the hghest point of land east of easting 60?
2. Give TWO reasons why there should be a flour mill at 603030.
3. Locate by means of grid references and name (a) a windmill, (b) a church with a tower, located west of the Great Ouse, and (c) the chapel in a village which has roads on both sides of a drainage channel.
4. What is (a) the distance apart in a straight line, and (b) along the nearest route by way of metalled roads, from Downham Market railway station to the level-crossing at 606070?
5. What does the map tell you of the meaning of the following words: Lode, Drove, Sluice, Wash, Dike?

**Photograph Question**
Describe the scene looking south across Denver Sluice. Consider relief, drainage, field pattern, and land use.

**Exercise B**
1. Draw a sketch map of the whole area shown on the map extract but reduce the scale to half that of the original. Mark on the map where you think the boundaries occur between fen silts, clay, and Greensand (refer to the cross section); label each area. Also mark on the Great Ouse, main drainage channels, Downham Market, all the villages, railways, and main roads.
2. What measures have been taken to improve drainage and prevent flooding in this area?
3. Analyse the site and shape of Downham Market and account for its importance.
4. Compare the route followed by the north–south railway with that of the A10(T) between the northern and southern edges of the map. Give reasons for the differences.
5. Until recently the A10(T) followed the course of the present B1507. What arguments would have been put for and against the construction of the new section of the A10(T) from 627054 south to 617014?

**Essay Work**
Draw a sketch map of the area around the Wash, marking in the principal rivers and the limit of the Fens. Mark and name FIVE important towns in or at the edge of the Fens.

Write a short essay on farming in the Fens explaining why the area is one of the most intensively cultivated arable areas in the British Isles and why there is very little livestock farming.

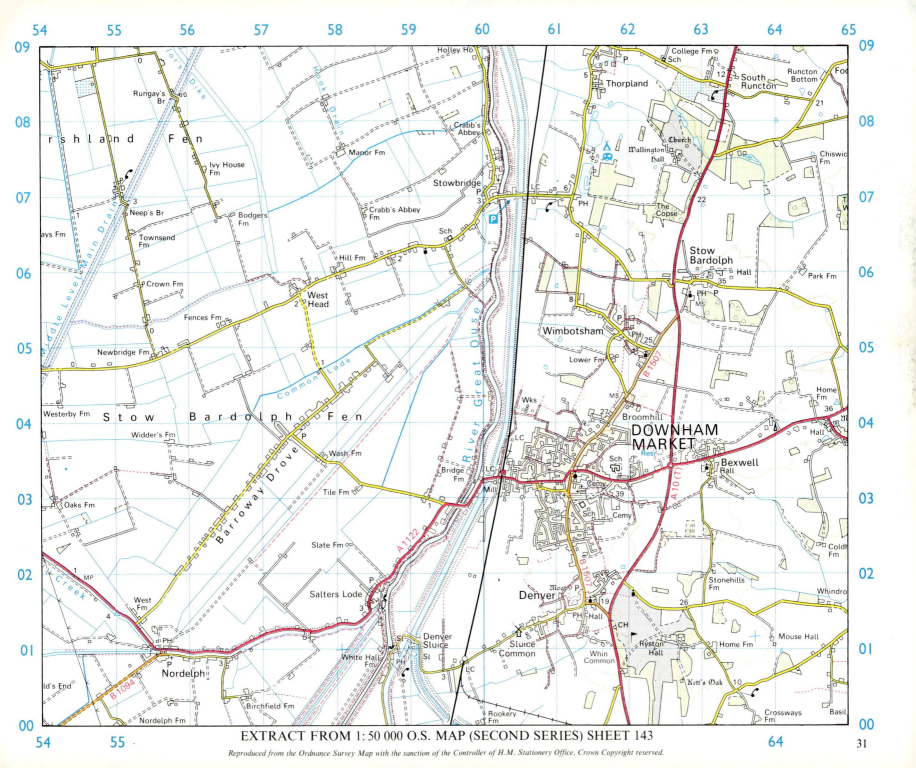

## THE UPPER TYWI AREA OF CENTRAL WALES

**River Development:** On impermeable uplands, rainwater will soon saturate the thin soils. Where there is a slope, numerous minute rivulets, perhaps a few centimetres wide, will trace intricate courses winding among the tufts of grass and heather. As these join up to form a stream of fast flowing water, a small notch is cut into the hillside, but this is not sufficient to show up on a map. In this *upland* or *moorland course*, the stream carries away small particles of soil and grains of rock. As other streams join, the river is eventually sufficiently powerful to carry along small pebbles and at once its erosive power is increased enormously. This is because rock fragments carried along the floor provide the 'teeth' which wear the stream bed down. Where the bed is uneven, rock fragments may scour out pot holes which pit the beds of most mountain streams and assist in lowering them further.

A river's erosive potential is influenced by its volume, by the slope of the land and therefore the rate of flow, and the load it is carrying. However, erosion is irregular, being most marked in times of flood when boulders several tonnes in weight can be moved.

As a river cuts downwards, its banks will be subject to weathering. Soil creep down the sides of the small valley is one process; another is called freeze–thaw which occurs in winter and leads to the shattering of exposed rock faces as moisture percolates into small fissures during the day and freezes and expands at night. Thus the valley is opened up into a narrow 'v-shaped' cross section, but this shape will depend very much on the resistance of the rock. If it is very resistant, there may be precipitous faces, broken only where tributary streams join the main one. The stage in river development we have considered in this paragraph is known as that of *youth*.

Where a river is no longer cutting downwards, a broader valley is found. The river current is strongest on the outside of a bend or meander, and so through lateral (sideways) erosion the river cuts into higher land forming a river cliff. On the inside of the meander deposition builds up a slip-off slope. Meanders migrate downstream, extending the river cliffs and reworking the deposited material of the slip-off slopes. Thus a flood plain is formed. When the river floods alluvium is spread over the flood plain. A river valley at this stage of development is sometimes described as being *mature*.

Figure 33 shows the forms of river valley that are to be found on the Upper Tywi extract.

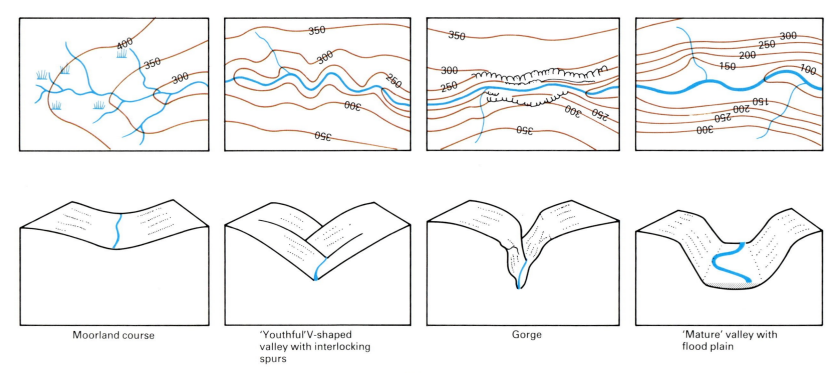

FIGURE 33
**River development in Central Wales**

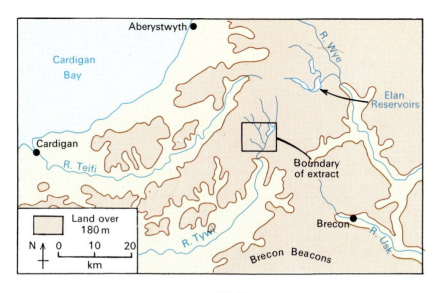

FIGURE 34
**Central Wales**

**The Upper Tywi Extract:** First locate the extract by studying figure 34. A section drawn across the extract would show that the upper surface between the valleys have a remarkably even level; in places these interfluves are narrow plateaux and in places they are more rounded, but in either case they rise to a height of just over 400 metres. This feature is apparent in much of Central Wales, where the ancient rocks of Ordovician and Silurian age have been worn down to a plain and then re-elevated. Three such erosion surfaces can be recognised as shown in figure 35. This extract is taken from a portion of the middle plateau which has been eaten into to form a greatly dissected landscape. The rock is impermeable, much of it being a slaty mudstone, and the rainfall is over 1800mm per annum so that processes of river erosion are very active. On the uplands where the land is level or slightly sloping, there are extensive peat bogs which reach a great depth locally where poorly drained hollows occur.

However, the present relief is not solely the result of river erosion on an uplifted plateau of ancient hard rock. There are traces of glacial deposits in some of the valleys, and although these do not have the spectacular glacial features found in mountainous areas such as Helvellyn, nevertheless there is no doubt that the main valleys on the extract have been deepened by glacial action.

**Forestry:** In recent years the Forestry Commission has added considerable tracts of this area to its afforestation schemes. The Sitka spruce responds better than most conifers to the ill-drained, very acid, marshy tracts on the plateaux above 350 metres, although some pine is found at these levels also. On the valley sides however a much wider range of species will tolerate the more sheltered, better drained terrain.

**Reservoirs:** Consider the following factors:
(1) Heavy rainfall. *What is the annual rainfall in this area?*
(2) Deep gorges. *Calculate how deep and how wide the most well-developed gorges are on this extract.*
(3) Impermeable rocks which have experienced little shattering. *What age and what type of rock is found here and what has shattering got to do with this topic?*
(4) Isolated, with a low density of population. *Estimate the density of population for this extract. What is the area of each grid square on the extract? Is the average density over 25, 10 to 25, 1 to 10 or under 1?*
(5) Increasing demands for water. *What is the nearest industrial region to this area?*
With these factors in mind it is not surprising that more reservoirs are being established in Central Wales. The River Tywi Scheme was officially opened in 1973; it involved the construction of Brianne Dam behind which are impounded the waters of the upper Tywi to form Llyn Brianne.

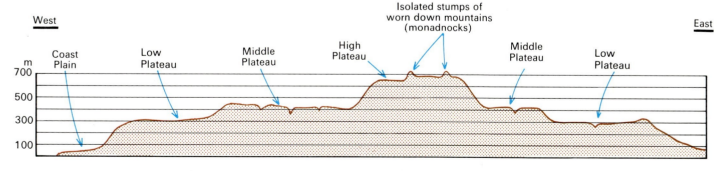

FIGURE 35
**Relief section across Central Wales**

# QUESTIONS ON THE UPPER TYWI MAP

**Exercise A**
1. Identify the types of vegetation at the following grid references:
   (a) 724522, (b) 721509, (c) 785505.
2. Pair the descriptions of the physical features with the relevant grid references:
   (i) a col, (ii) a gorge, (iii) a plateau; (a) 718465, (b) 779488, (c) 755505. Refer to page 5 for definitions of these terms.
3. Plan a hike along roads, tracks and footpaths which are marked on the map from the bridge at 764451 to the road at 717538. Work out precise instructions for the party leader to follow assuming that he will have both a map of the area and a compass. Divide the route you give into legs and for each leg point out how long it is and what the terrain is like. For example:
   Leg 2; distance 3km. Climb up from the west of the building at 763484; proceed northwards keeping the scar face to your left. Continue over rough pasture on the plateau top, travelling north westwards for 2km. Descend into a valley and pass building 744509 on your immediate right.

**Practical Work**
Enlarge an area of the map between eastings 75 and 78 and north of northing 50. Draw in contour lines and drainage courses and build this up into a three-dimensional model. A minimum workable size for the model is probably 15cm by 20cm.

**Exercise B**
1. Draw a sketch section from 730480 to 760500. Mark where the two water-courses cross the line of the section. In what ways do the two valleys differ?
2. Describe the relief and drainage of the area south of northing 47.
3. Relate the settlement and communications to the physical geography of the area.
4. Study the photographs on this page and then describe the features of the valley in each of the scenes. Locate a similar cross section of a valley for each of the illustrations on the map extract and give the relevant six-figure reference.

**Essay Work**
What factors must engineers consider when choosing a site for a reservoir and what problems, both physical and human, can they meet? How might these be overcome? Besides storing a supply of fresh water, what amenities might the reservoir and its surrounding area provide?

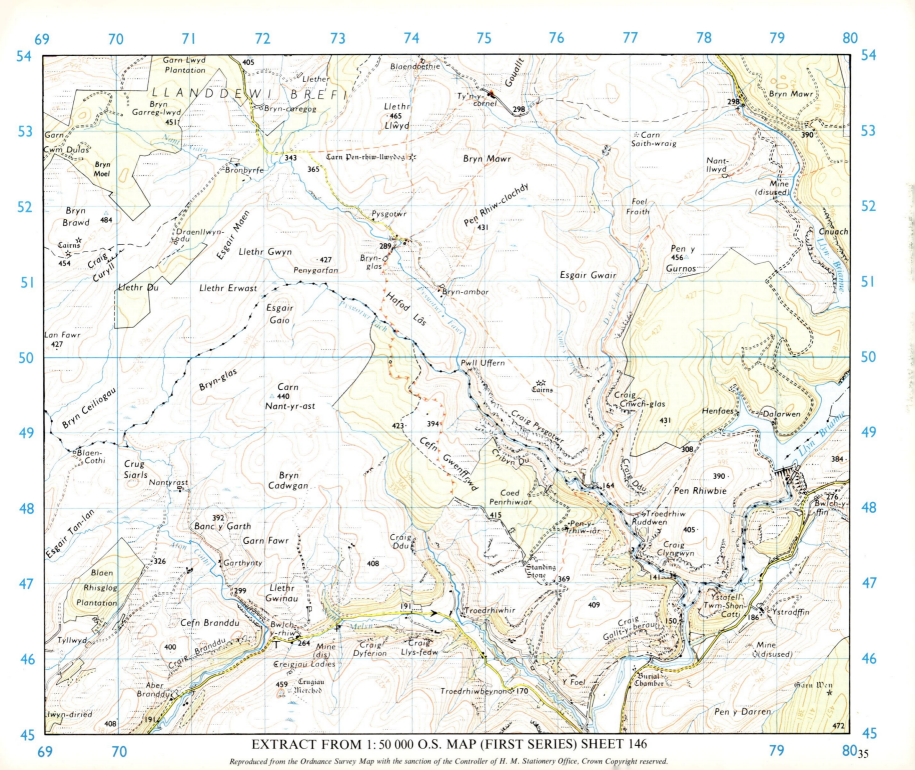
EXTRACT FROM 1:50 000 O.S. MAP (FIRST SERIES) SHEET 146

Reproduced from the Ordnance Survey Map with the sanction of the Controller of H. M. Stationery Office, Crown Copyright reserved.

# MILFORD HAVEN

For many years Britain was dependent upon coal to satisfy most of its energy requirements. However, the availability of different means of fuel and power supply (oil, natural gas and nuclear power) led to a relative decline in the importance of coal (see page 45). Throughout the 1960s and early 1970s large quantities of crude oil were imported, at an attractive price, from countries surrounding the Persian Gulf. At the same time a tremendous expansion took place in refining capacity. Coastal sites were usually chosen for the new refineries, as this was the place where the cargo had to be transferred from one form of transport to another, i.e. a break of bulk point. Sheltered areas of deep water were particularly sought after. One such area was Milford Haven.

**Milford Haven:** If you study the map on page 38 you will see that Milford Haven is the name given both to the estuary of the River Cleddau and the main town on its north bank. The estuary was at one time a river valley carved out of the underlying Old Red Sandstone and then submerged as the sea level rose. We call this feature a ria, and there are other examples to be found in south Devon and Cornwall.

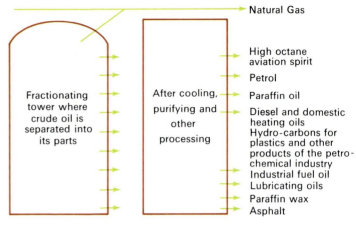

FIGURE 37
**The fractionating tower**

**The Refineries at Milford Haven:** Four oil companies established refineries at Milford Haven. In addition B.P. established a terminal to supply its refinery at Llandarcy, near Swansea. Besides a natural deep water channel, further dredging enabled tankers of 270 000 tonnes to berth at the jetties in the sheltered estuary. Land was available for development and the position to the west of the country suited tankers arriving from the Middle East as they did not need to pass through the congested English Channel. *Which of these locational factors would you not consider as an advantage now?*

Crude oil is of little use in an unrefined state. The function of the refinery is to separate out the different products that we have come to regard as essential to our way of life – petrol, lubricating oils, and diesel fuel. This it does in a fractionating tower where the crude oil is boiled. The lightest materials are removed at the top of the tower, with the heavier fractions being drawn off as liquid side streams.

A small percentage of the refined oil from Milford Haven travels by road and rail. The majority is re-exported in smaller tankers for distribution around our coasts, or piped directly to the Midlands and North-West of England. Some is retained for use in the nearby electricity generating station.

**The Conflict Between Industrial Development and Conservation:** The refineries at Milford Haven straddle the boundary of the Pembrokeshire Coast National Park. Although employment is provided for many workers, conservationists are critical of the effect those refineries have on this area of outstanding natural beauty. The oil companies themselves have spent a large amount of money in reshaping the land so that the refineries are hidden from view. In combating pollution they have applied noise control to machinery, run effluent through oil filters to ensure the purity of the estuary, and built tall chimneys to release combustible gases high into the air.

*Exercise:* Make a list of all the National Parks in Britain. Does a conflict between industry and conservation occur in any of the areas you have listed? Under what conditions should industry be allowed in areas designated as places of natural beauty?

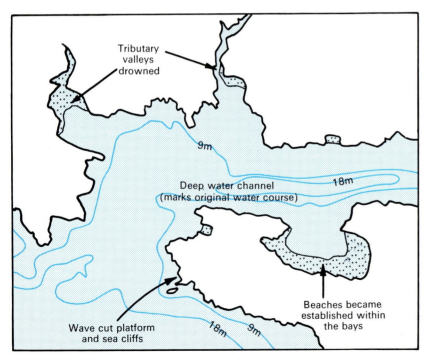

FIGURE 36
**Milford Haven ria**

# HIGH MARNHAM

**The Trent Vale:** From Burton-on-Trent in Staffordshire to the Humber, the River Trent takes a meandering course within a flood plain which is generally not much wider than three kilometres. The gradient of the river throughout this stretch of its course is very gentle so that it is unable to continue deepening its channel any further. Instead it winds its way from side to side across the flood plain, occasionally impinging against solid rocks and there forming steep river cliffs known as bluffs. The meander loops may become more accentuated in time, until eventually the meander is cut off and the course of the river straightened. Portions of the abandoned course of the river can often be detected on a map; sometimes for example a district or county boundary follows the abandoned course of the river rather than the present one.

Settlement established before the present century was usually absent from the flood plain, but a line of villages on either side have been established on the firm founations of terraces. These terraces may well represent a former, wider flood plain of the Trent which has subsequently been rejuvenated, enabling it to erode its present flood plain. Nowadays embankments afford protection from flooding, and the flat valley floor can be settled on. The river is navigable downstream from Nottingham for barges of up to 200 tonnes and the main cargoes carried include petroleum, grain, timber, gravel, cement, and coal.

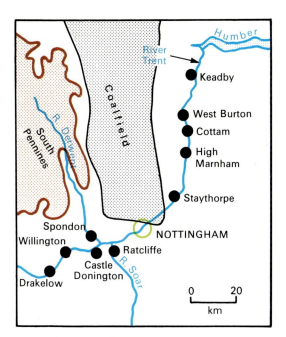

FIGURE 39
**The Trent power complex**

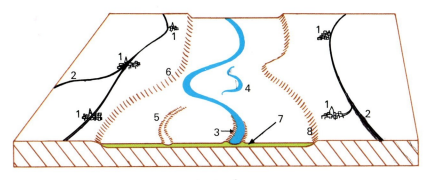

FIGURE 38
**The Trent flood plain below Newark**

*Exercise:*
(a) Comment on the siting of the villages (1) and the courses of the roads (2).
(b) Identify: river bluff; embankment; abandoned course of river; partially abandoned course, terraces, and alluvium. These are numbered 3 to 8, but not in that order.

**Power Stations:** Electricity can be produced by using many sources, including water power, nuclear energy, coal and oil. In this country by far the most important is coal. Similarly, the electric power stations provide the Coal Board with its biggest market. The largest single concentration of generating capacity in Britain lies along the banks of the River Trent. All these stations, located in figure 39, are of the conventional thermal type, using coal as the fuel. With cooling towers rising to over 100 metres and chimneys approaching 200 metres high, their physical impact is enormous. In terms of output these stations together produce approximately 20 per cent of England's electricity, much of which is transmitted to the south-east of the country. The idea of a power station catering for local demand is outdated. These power stations were located in the Trent Valley because:
(1) It is centrally situated in relation to other parts of the country.
(2) There is flat land available for construction.
(3) The fuel is easily obtainable from the nearby largest and most modernised coalfield in Britain.
(4) Water from the river can be used for the cooling process.
(5) There are facilities for the disposal of the ash from the boilers; these include its usage in the reclamation of gravel pits and the building up of low-lying land.

**High Marnham:** When it was open in 1962, High Marnham Power Station was Britain's first one million kilowatt power station. At full load, 10 000 tonnes of East Midland's coal is used each day. This is pulverised, then carried by hot air to the boilers where it heats the purified water. The steam, superheated and at high pressure, is able to drive the turbines which control the generators that will produce electricity. The steam meanwhile passes through a condenser. The water used to cool the steam becomes warm itself and is taken to the cooling towers and then returned to the condenser. Over 120 million litres of water are required every hour for cooling, although much of this can be used again and again. The energy produced at High Marnham is fed into the country's Super Grid.

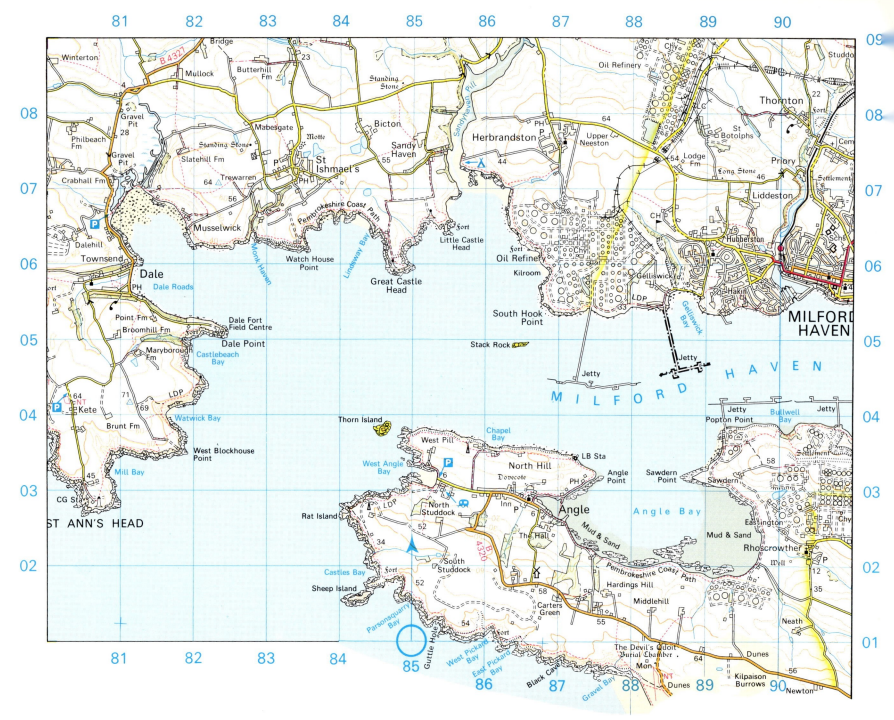

EXTRACT FROM 1:50 000 O.S. MAP (SECOND SERIES) SHEET 157

Reproduced from the Ordnance Survey Map with the sanction of the Controller of H.M. Stationery Office, Crown Copyright reserved.

# QUESTIONS ON THE HIGH MARNHAM AND MILFORD HAVEN MAPS

### Exercise A
1. Comment on the position of (a) a lighthouse at 807028 and (b) a jetty in square 8704 (MH).
2. What direction is the triangulation pillar (823071) from Angle Church (866029) (MH)?
3. Suggest what is extracted from the flood plain of the River Trent.
4. The following statements may be true or false. Study the map extract of High Marnham to find out:
   (a) Low Marnham (807695) is a good example of a linear-shaped village.
   (b) The Trent forms part of the boundary between two districts.
   (c) There are many roads crossing the Trent.
   (d) There is no woodland to the east of the Trent.
5. Give THREE pieces of evidence of prehistoric settlement on the Milford Haven map.

### Practical Work
Draw a map on the same scale as the High Marnham extract. Mark on it the Trent and its tributaries; draw and label the course of the A1133, and mark and label the position of the main settlements.

Describe the sites and positions of these settlements and the route taken by the A1133.

### Exercise B
1. Describe and account for the distribution of industry on the Milford Haven extract.
2. Compare the coastal features from Sheep Island (843017) to Gravel Bay (878006) with those from Dale Point (825052) to Monk Haven (828064) (MH).
3. Explain why there is little settlement south of the B4320 (MH).
4. Describe the course of the River Trent and the form of its flood plain.

### Essay Work
Give as many examples as you can of recent industrial development which is to some degree dependent upon a site close to water. Describe and illustrate with sketch maps the various examples you use.

left: EXTRACT FROM 1:50 000 O.S. MAP (FIRST SERIES) SHEET 157
right: EXTRACT FROM 1:50 000 O.S. MAP (SECOND SERIES) SHEET 121

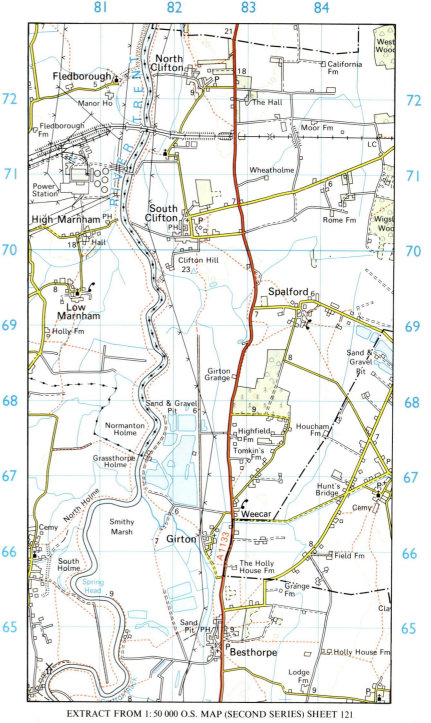

EXTRACT FROM 1:50 000 O.S. MAP (SECOND SERIES) SHEET 121

*Reproduced from the Ordnance Survey Map with the sanction of the Controller of H.M. Stationery Office. Crown Copyright reserved.*

# HELVELLYN

During the Ice Age, glaciers and ice sheets covered much of the British Isles as far south as the Thames valley and lower Severn. Large areas of the lowlands were influenced by glacial deposition including boulder clay, sands and gravels left by the ice sheets. It is in the mountain regions, however, that we find the most striking evidence of glaciation, especially in the Highlands of Scotland, Snowdonia and the Lake District. Many features of upland glaciation are particularly well illustrated in the Helvellyn area.

Helvellyn is one of the well-known peaks in the Lake District. The rocks making up the area covered by the map extract belong to the Borrowdale Volcanic Series, and are a consolidated mass of varied lavas and ashes which have been eroded to give very rugged scenery. The contour patterns on the map extract probably appear very confusing at first sight. In a case such as this you should first locate the highest stretches of land, then find where the lowest areas of land are, and finally work out which are spurs projecting from the highland and which are the valleys going up into the highland, watercourses being the obvious clue in the latter case.

Glaciation has been the major influence on the details of relief. During the Ice Age, an ice cap covered much of the Lake District, although its limits must have fluctuated considerably. Tongues of ice, known as glaciers, moved outwards from the centre following the courses of the main river valleys. The ice plucked away fragments of rock and this debris then became embedded in it, providing the glacier with 'teeth' to further erode its valley; this last process is known as abrasion. Thus the valleys were deepened and straightened. The lower slopes of the western portion of the Helvellyn mass form the side of such a glaciated valley. Large ribbon lakes occupy hollows in the main valleys. The hollows were scooped out by glacial erosion, and sometimes their lower ends were blocked by glacial deposits. Thirlmere is an example of such a lake. It has supplied water to Manchester since 1894. The water level of the lake has been raised artificially and the outflow controlled. Coniferous plantations have been established on its banks in order to conserve the slopes from excessive erosion and so prevent silting up.

In the later stages of the Ice Age, when much of the ice cap had decayed, snow still accumulated, often in hollows high up on the mountain sides. During the day some of the snow round the edges of such hollows may have melted and the water penetrated cracks in the rock. Freezing of this water at night could cause rock shattering. Also, as snow accumulated, the under layers would be pressurized to form ice which, in turn, would pluck away rock fragments. These processes deepened the hollows to form cirques (also known as corries and cwms). These are semicircular basins often with an almost vertical face behind them. The ice may have finished at the lip of the cirque where a small terminal moraine would then form. Behind this today a circular lake or tarn often occupies the floor of the hollow. When two cirques eat back towards one another, the intervening upland can be worn into a sharp-edged ridge called an arete; Striding Edge is a fine example of this.

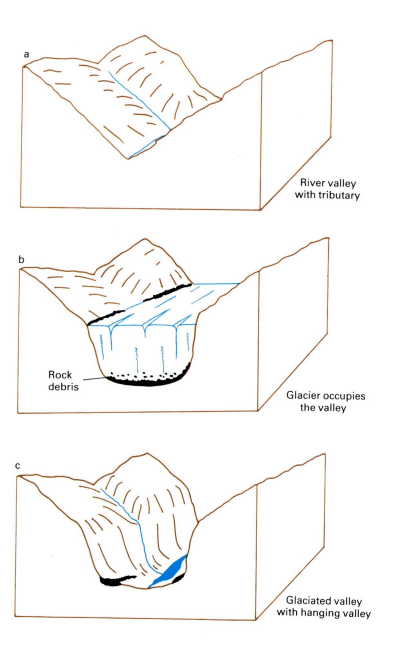

FIGURE 40
**Glaciation of a river valley**

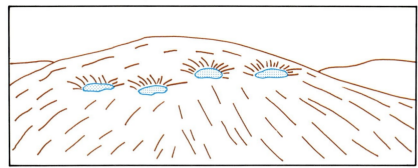
(1) Hollows with accumulating snow and ice

(2) Hollows being deepened into cirques or corries: tongues of ice leading from the hollows

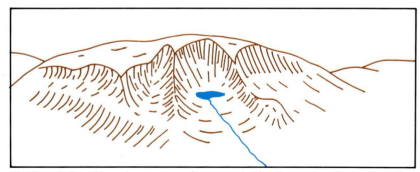
(3) After glacial action: cirques or corries separated by sharp edges known as aretes

FIGURE 41
**Development of a glaciated landscape**

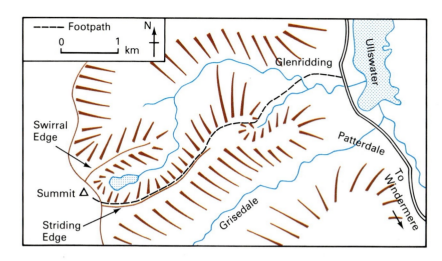

FIGURE 42
**The course of a footpath to Helvellyn summit**

**Outdoor Leisure Map Series:** For many areas which are frequently visited, special editions of the 1:25 000 maps known as the Outdoor Leisure Map Series, are available. Extra information of particular value for walkers is included. Figure 42 shows the course of a very popular walk to the summit of Helvellyn.

*Exercise:* What information is provided on the Helvellyn extract from the English Lakes Outdoor Leisure Maps that would be of special interest to a fell walker?

## QUESTIONS ON THE HELVELLYN MAP

**Exercise A**
1. What is the character of the land surface at 350160 and 354159?
2. Calculate the height of Red Tarn and Thirlmere above sea level.
3. Give directions to a walker who wishes to climb Helvellyn from the main road, keeping to marked footpaths. He has a compass but no map.
4. Give an example and explain what is meant by each of the following terms: Pike; Crag; Gill; Beck.

**Practical Work**
On a grid showing all the eastings and northings, mark in the 350, 500, 650, 800 and 950 metres contours. Colour the layers green (the lowest land), yellow, orange, light brown, dark brown and black respectively. Print the words CIRQUE, ARETE, TARN and SUMMIT in the correct places.
  If you wish, you can make a layer model using the map you have made.

**Exercise B**
1. Draw a sketch section along a line from 340170 to 350140; mark and name on it any features that you associate with glaciation. Explain the origin of these features.
2. Study the course and valley of Nethermostcove Beck from its source to its junction with Grisedale Beck. Describe and give reasons for the shape of its valley at 346147, 348144, 353143 and 358144.
3. Why does the lake shore protrude slightly at 323137?
4. Comment on the distribution of woodland.

**Essay Work**
Why is the Lake District such a popular tourist centre?

**Photograph Question**
Make sketches of the scenes shown in the photographs and then, using information given on the map extract, label as many features as possible.

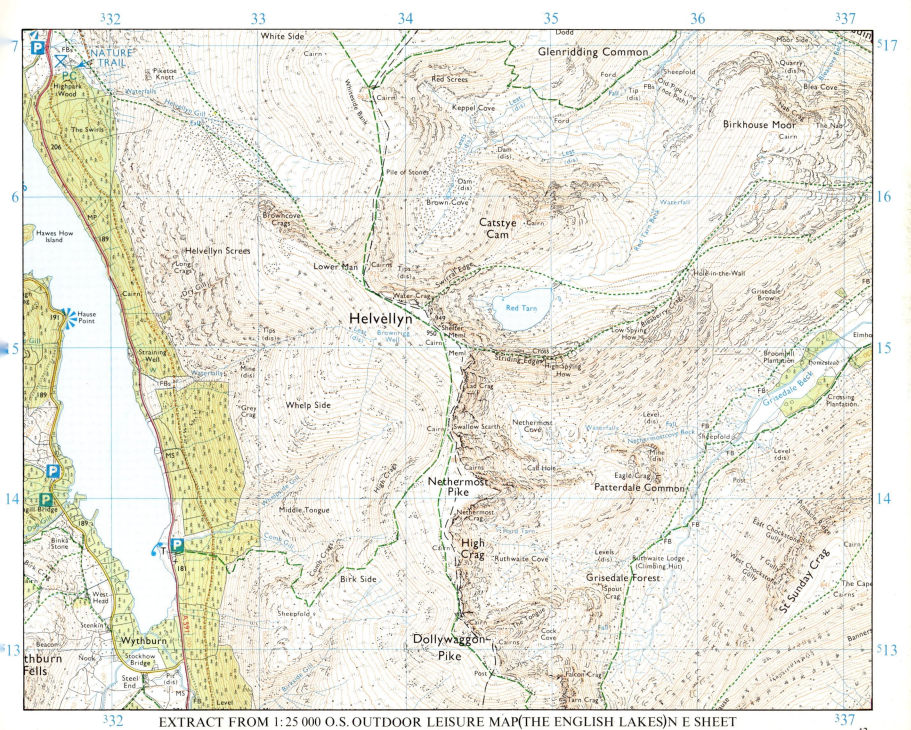
EXTRACT FROM 1:25 000 O.S. OUTDOOR LEISURE MAP (THE ENGLISH LAKES) N E SHEET
Reproduced from the Ordnance Survey Map with the sanction of the Controller of H.M. Stationery Office, Crown Copyright reserved.

# THE RHONDDA

**The South Wales Region:** South Wales is sometimes recognised merely as a coal-mining area. Although its industrial development has in the past been based on coal, even this is not true today. Much of South Wales has not been industrialised at all. The fertile coastal lowland composed of marls, sandstone and limestone (known as the Vale of Glamorgan) still retains much of its rural character. In the north of the Vale this situation is changing as the presence of the main lines of communication between Swansea and Cardiff has encouraged industrial development. This growth has been further stimulated by the movement of people from the coalfield as the industries now providing employment are freer to choose more accessible positions than the coal-based industries of the past.

To the west of Swansea lies the Gower peninsula. Similar to the Vale of Glamorgan, it is essentially an agricultural area that has become a dormitory and recreation area for the nearby urban region, as well as a summer resort for visitors.

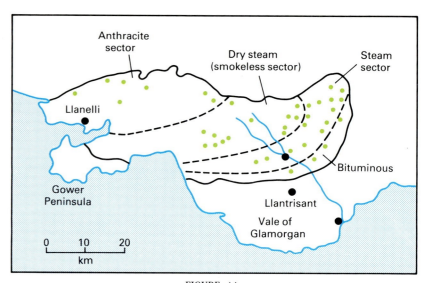

FIGURE 44
**Distribution of collieries on the South Wales coalfield. The Rhondda Fawr and Fach have been inserted.**

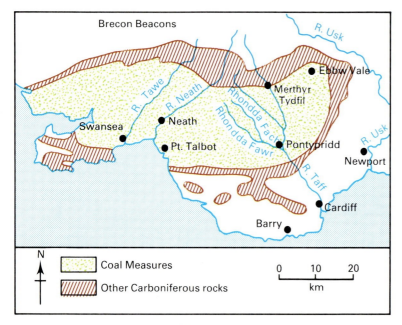

FIGURE 43
**South Wales**

From the industrial viewpoint the most significant of the region's geological features has been the coal basin. The Coal Measures of South Wales are classified into the Lower Coal Series, the Pennant Sandstone Series and the Upper Coal Series. The Pennant Sandstone forms a barren upland area that rises from 300m in the south to just under 600m in the north. Although the rock has been subjected to folding and faulting, the Pennant Sandstone appears as a dissected plateau continuous with the high plateau of Central Wales. This suggests that these surfaces have been planed by agents of erosion, though whether the principal agent has been the sea or rivers is still a matter for discussion.

Streams have cut deeply into the sandstone plateau, forming steep-sided valleys. These valleys became far more important economically than the surrounding upland. Since the valleys were incised into the Upper Coal Measures it became possible, during the early stages of mining, to run tunnels or adits laterally into the sides along the coal seams. Today, however, most coal is obtained from underlying beds and brought to the surface by way of shafts.

South Wales is remarkable for the variety of types of coal, ranging from bituminous to steam, dry steam and anthracite. Dry steam and anthracite are classified as naturally smokeless fuel under the Clean Air Act. Welsh anthracite contains up to 95 per cent carbon, giving it a high calorific value and low ash content.

In recent years a massive investment programme in the coal industry has led to the modernisation of some of the long-life pits, most of which lie in the east of the coalfield. The first new mine for ten years was opened in 1978, near Ammanford in the west of the area. This was unusual because the trend has been for mines to close and jobs to disappear. Competition from alternative fuels and a recession in manufacturing, especially in the steel industry, have meant that only the most economic pits have remained open.

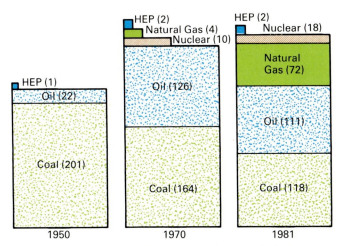

FIGURE 45
**Sources of energy in Great Britain. Comment on the change which took place between 1950 and 1981.**

Post-war governments have been concerned with trying to reduce unemployment in South Wales. Development Area status, industrial estates and now the Lower Swansea Valley Enterprise Zone aim to lure businesses to the region. Many firms have come already to the coastal lowlands, including the recent Ford plant at Bridgend.

More jobs are still required and local authorities continue to emphasise the benefits of a location in South Wales. Financial incentives such as grants and low factory rents can be found elsewhere. However, a southerly location in Britain combined with the improved accessibility of the region is a positive attraction.

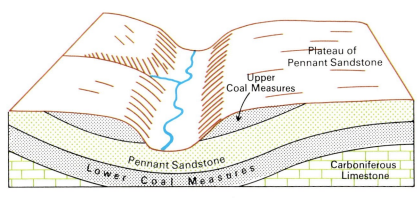

FIGURE 46
**Section across the Rhondda Valley**

**The Rhondda:** The Rhondda is not a town but the name given to two valleys, the Rhondda Fawr and the Rhondda Fach (fawr–great; fach–little). There are a number of important tributary valleys, one of which, Cwm Parc, appears on the extract. This area was unaffected by the early stages of the industrial revolution in South Wales. Whilst iron smelting was developing in the eastern coalfield, the Rhondda area remained sparsely populated. A few farmers used the valleys mainly for grazing sheep. By the early 1860s the Rhondda began to feel the impact of industrialisation. The change was rapid. It was then that the railways were expanding, and at the same time Britain's trade grew, as did her merchant fleet. Both required large quantities of coal for raising steam, and the Rhondda coal had a reputation for being the finest coal in the British Isles for this purpose.

Thus the Rhondda valleys became corridors of industrial development. In the valleys were cramped pitheads, housing and railway sidings. Often the sites for long lines of terraced houses were cut out of the valley sides. Today, these houses, built of pennant sandstone, stretch for miles as village has merged into village. The map extract shows the close relationship between the valleys and the built-up areas. In contrast, the plateau remains bleak and bare. The Rhondda, in common with most mining districts in the British Isles, continued to expand until just before the First World War. In 1924 some 40 000 men in the Rhondda were employed in about 40 collieries. The economic depression of the twenties and thirties hit this region hard, with unemployment rising to between 40 and 50 per cent. As the mining industry recovered, it was faced with severe competition from petroleum. Railways turned to diesel fuel and electricity, and shipping to heavy oils; hence the demand for steam coal was enormously reduced. Only one colliery now remains open in the Rhondda, at Mardy.

In order to alleviate the unemployment that has resulted from these changing circumstances, attempts have been made to bring new industry to the Rhondda, but these have met with only partial success. Building sites are limited, and the coastal plain offers more scope for development. Consequently many workers have to commute to other areas for employment. Future job opportunities for the inhabitants of the Rhondda seem to depend on the industrial development taking place in the Vale of Glamorgan. This will either encourage the settlements in the valleys to perform a dormitory function, or eventually will lead to a migration of people from the area. As the Forestry Commission continues to take over the open moorland, and as old colliery workings and waste heaps are cleared, the valley sides seem destined to become pleasant forested areas.

## QUESTIONS ON THE RHONDDA VALLEY MAP

### Exercise A
1. What is the distance from the road junction at 957962 to the road junction at 938985 (a) in a straight line, and (b) by road?
2. Which of the following types of slope – concave, convex, uniform – is to be found at (a) 953978, (b) 925960, (c) 970972?
3. What is the vegetation cover at (a) 929962 (b) 922972 (c) 953967 (d) 945999?
4. Draw a sketch section from 924980 to 953997 and mark in and label as many different man-made features as possible along the line of the section.

### Exercise B
1. Describe the valley of Cwm Parc from its head to 940957.
2. Account for the distribution of settlement on the map.
3. What different types of evidence are there of present and former mining activities in the area. Give grid references of the evidence in your answer?
4. Compare the route taken by the A4061 to the north of the junction at 938985 with that to the south.
5. Describe and account for the location of factories in the Rhondda.

### Essay Work
Discuss the problems associated with coal mining in the 'valleys' of South Wales.

### Photograph Question
Compare the two scenes shown by the photographs with respect to their physical and human geography.

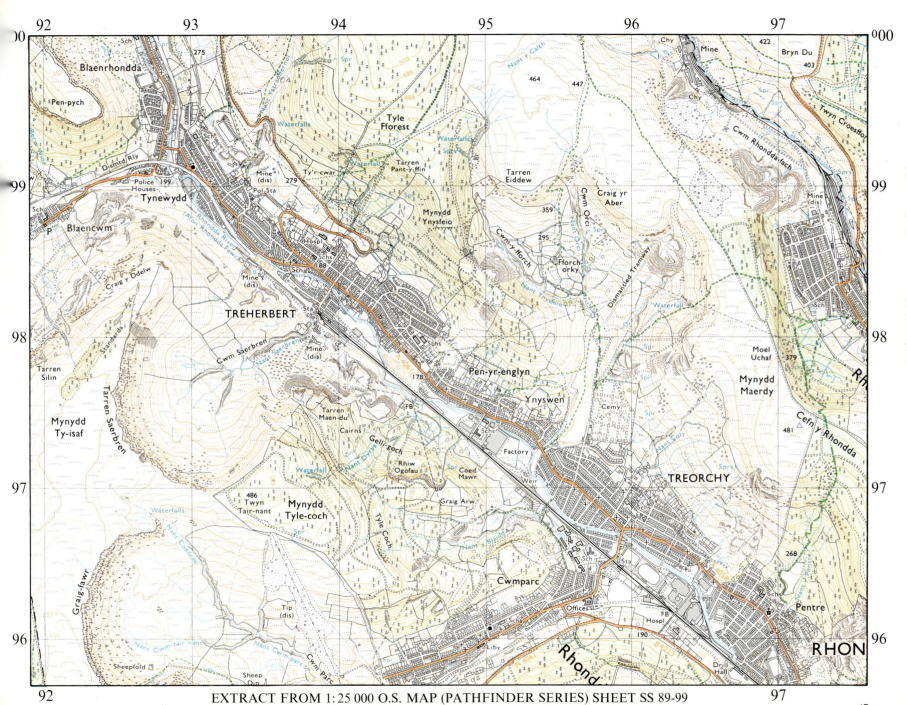
EXTRACT FROM 1:25 000 O.S. MAP (PATHFINDER SERIES) SHEET SS 89-99
Reproduced from the Ordnance Survey Map with the sanction of the Controller of H.M. Stationery Office, Crown Copyright reserved.

# CAMBRIDGESHIRE – VILLAGE STUDIES

In the introductory section on settlement, attention was given to site, situation, function, and pattern of rural settlement. You should refer again to that section (page 11) before considering the features of settlement on the next map extract.

The villages on the map extract lie immediately to the north-west of Cambridge. A close inspection of the map, and the accompanying diagram showing the geology, indicates that although some dispersed settlement occurs in the fen area, the fenline itself is marked by a chain of villages including Cottenham, Rampton, Willingham, and Over. These villages, with Anglo-Saxon place name elements, preferred the sounder foundations that a site on Greensand or Ampthill Clay could offer to that of a site on the fen. Another chain of settlements can be distinguished further south, this time following a course parallel with the Roman road, the Via Devana. This group includes Histon and Impington, villages that are rapidly developing into dormitory settlements of nearby Cambridge.

The shapes of these villages vary considerably. Shape is influenced by the site and position of the settlement, the relative importance of defence, and the agricultural systems that have been practised. Many of our rural settlements have developed under the influence of a road network.

**Street Villages:** In this case the buildings stand side by side along a single highway, as in Boxworth End, Swavesey. Other linear villages may have been built along river terraces, drainage canals, and along a line of springs; although such settlements spread along a roadway today, they should not be thought of as street villages.

**Crossroad Villages:** This is where buildings are strung along more than one road. You expect to find the church and local shops at the junction of the major routeways.

Village shapes are not always determined by existing road patterns.

**Green Villages:** These are very distinctive forms. They probably originated as forest clearings in Anglo-Saxon times and then the greens became useful as secure places for pasturing cattle at night. Both Rampton and Willingham have greens, although this is not apparent on the map extract. These greens are also the site of the village pump, making them ever more the focal point of the settlement.

In the past the inhabitants of most villages were primarily engaged in agriculture. In this area of rural Cambridgeshire orchards have thrived on the loamy soils which are the result of the overspreading of sandy gravels upon clay. Small fruits and vegetables are also cultivated on smallholdings and provide a supply of fresh produce for the markets of our major cities or, alternatively for food processing factories. However, in common with many other areas which had once a wholly rural character, this area has acquired many features of settlement which are remote from agriculture. Recent years have seen the emergence of the dormitory village, or the village now dominated by its large dormitory estate. With increasing mechanisation of farming, resulting in the reduction in the number of farm workers, and improved communication links between town and village, what characterised the 'village community' is fast disappearing.

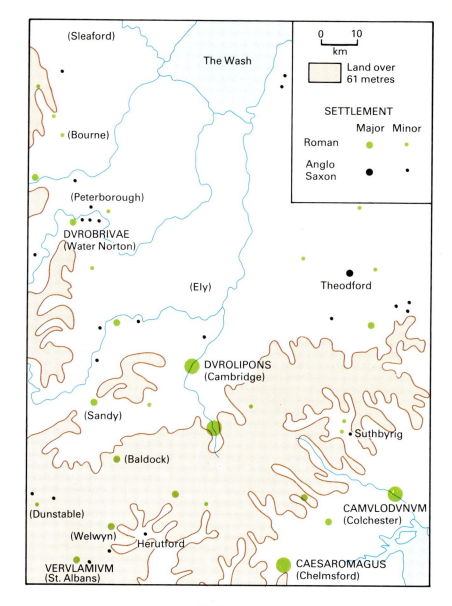

FIGURE 47
**Distribution of settlement over a portion of eastern England in Roman and in Anglo-Saxon times. The present day names of some of the settlements are in brackets.**
*Compare this settlement pattern with that on a modern map in your atlas. Point out the similarities and the differences and account for these.*

# CORRELATION OF SETTLEMENT AND GEOLOGY

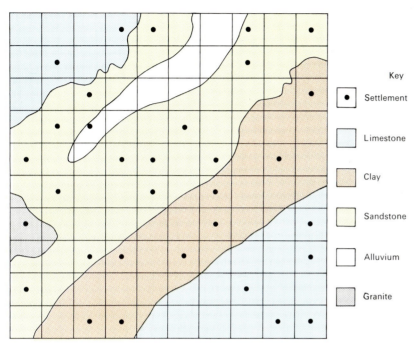

FIGURE 48
**Settlement and geology**

The correlation of the distribution of rural settlement with variables such as geology and relief by the use of matrices (or grids) can provide a more accurate picture than descriptive observation. In the following hypothetical example the aim is to correlate the relationship of settlement with surface geology.

**Matrix 1** (Observed data)

| Surface Geology | Settlement | Non-settlement | Total |
|---|---|---|---|
| Limestone | 7 | 18 | 25 |
| Sandstone | 15 | 23 | 38 |
| Clay | 8 | 19 | 27 |
| Alluvium | 1 | 7 | 8 |
| Granite | 1 | 1 | 2 |
| Total | 32 | 68 | 100 |

Matrix 1 tabulates the number of squares dominated by each type of surface geology, and the number of squares with and without settlement. We see that 32 of the 100 squares contain settlement. If settlement were associated equally with all rock types then it would be reasonable to assume that 32/100 of each type of surface geology would have settlement. Thus limestone which occurs in 25 squares should have 25 × 32/100, that is 8 settlements. A calculation like this can be made for each part of the matrix and a new matrix produced which shows the expected situation if each type of geology has its proper share of settlement.

**Matrix 2** (Expected distribution)

| Surface Geology | Settlement | Non-settlement | Total |
|---|---|---|---|
| Limestone | 8 | 17 | 25 |
| Sandstone | 12 | 26 | 38 |
| Clay | 9 | 18 | 27 |
| Alluvium | 3 | 5 | 8 |
| Granite | 0 | 2 | 2 |
| Total | 32 | 68 | 100 |

Compare Matrix 1 and Matrix 2 and the associations between settlement and geology are quickly spotted. For example, instead of 12 settlements on sandstone there are in fact 15. A useful way to tabulate these associations is to divide the actual observations by the expected distribution and then to use the answers to complete Matrix 3.

**Matrix 3**

| Surface Geology | Settlement | Non-settlement |
|---|---|---|
| Limestone | 0·9 | 1·0 |
| Sandstone | 1·2 | 0·9 |
| Clay | 0·9 | 1·0 |
| Alluvium | 0·3 | 1·4 |
| Granite | — | 0·5 |

Considering Matrix 3 we can see the association between settlement and sandstone is stronger than for any other rock type. Limestone and clay have an average association, whilst that between settlement and alluvium and granite is quite weak.

This analysis can be extended to correlate any patterns of distribution. We have considered settlement and geology. Land use and relief could also be correlated in a map analysis. Using calculations like this it is possible to make more precise conclusions when interpreting the information provided by such detailed maps.

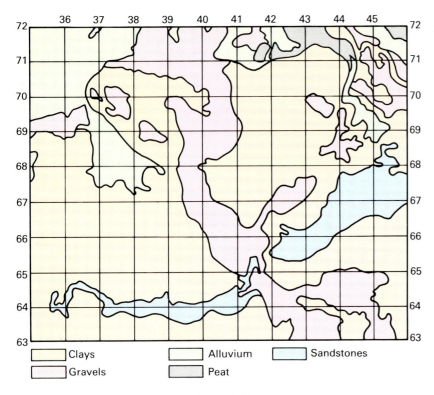

**FIGURE 49**
**Geology of the Cambridgeshire extract**

### Exercises on the Surface Geology Map
1. Construct a matrix, as shown in the example on page 49, for the OBSERVED relationship between settlement and surface geology. (For each square decide which is the dominant surface geology and whether any settlement at all is present.)
2. Construct a matrix for the EXPECTED distribution and compare this with the observed matrix. Comment on the relationships between clay, gravel, alluvium and settlement.

## QUESTIONS ON THE CAMBRIDGESHIRE MAP

### Exercise A
1. Name THREE villages that are possibly of Anglo-Saxon origin. Refer to page 9.
2. In grid squares 3569 and 3868, the railway has to cross physical obstacles and minor roads. Draw labelled plans to show how the railway copes with these obstacles.
3. A stranger wishes to get to Over church (372708) from Lolworth (366640). What instructions might you give him on the easiest route to take?
4. Identify the symbols at 374644, 452637, 426682, 356657 and 366642.
5. Give as many different pieces of evidence as you can find, quoting reference points, which show how drainage has been improved in this area.

### Exercise B
1. On a sketch map make a division of the area of the map into three physical regions. Relief may help you, but refer to the geology map as well. Describe the differences between each area.
2. Suggest what different forms of agriculture are practised in the area. Give reasons for your answers.
3. Describe the road patterns of Cottenham and Histon. Refer back to the section on Hull for help on this. What do the descriptions you give suggest about the character of these villages?
4. Calculate the distances between the nucleated settlements on the map. What pattern emerges from these measurements and why should this be so? Refer to page 11.

### Essay Work
Compare and contrast this area of rural Cambridgeshire with the area shown on the Downham Market sheet.

### Practical Work
Draw plans of Swavesey, Rampton and Oakington. Describe the contrasts in the forms, and the similarities in the functions, of the three villages.

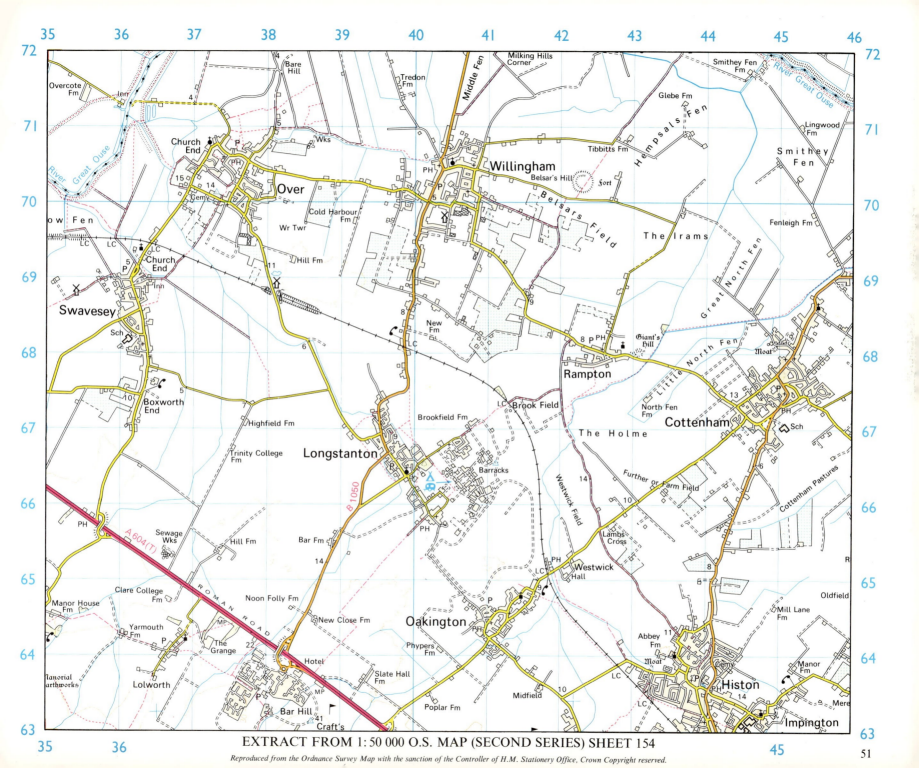
EXTRACT FROM 1:50 000 O.S. MAP (SECOND SERIES) SHEET 154
Reproduced from the Ordnance Survey Map with the sanction of the Controller of H.M. Stationery Office, Crown Copyright reserved.

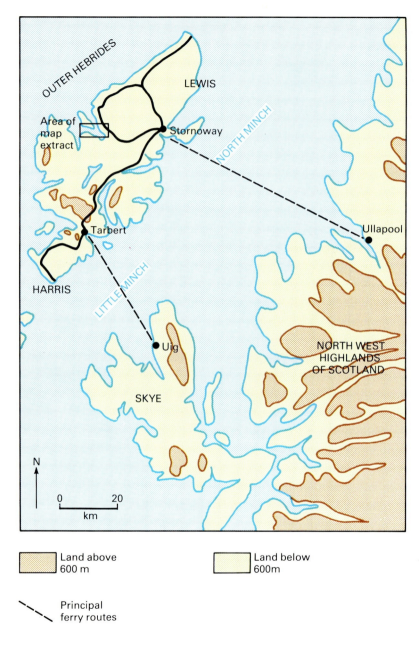

FIGURE 50

# ISLE OF LEWIS

Lewis is the largest of the islands that form the group known as the Western Isles or Outer Hebrides. The islands were formerly the summits of a mountain ridge, which was submerged by the sea to produce an island chain that stretches for over 200 kilometres. The sea passage that separates the Western Isles from the mainland of Scotland is the Minch. This is well known for its stormy nature, particularly in winter and partly accounts for the relative isolation of the islands.

**Scenery:** Lewis has a remarkable landscape. The main rock type is gneiss, which was formed in Pre-Cambrian times and is metamorphic in origin. Long periods of weathering and erosion by water and ice have laid bare the surface of the gneiss and produced numerous indentations, giving rise to a chaotic pattern of hills and valleys. A semi-aquatic appearance has emerged as hollows have filled with water. These lochs are dotted over the knobbly surface, which rarely rises above 200 metres in the north of the island. There is little evidence of deposition by ice action, but the decomposition of moss and heather has led to peat forming in depressions.

This 'knock and lochan' landscape is a good illustration of the effects of ice in lowland areas and is in contrast to the upland glacial features seen in the granite region of south Lewis and Harris, or in the English Lake District. *Compare the relief and drainage of the Isle of Lewis map extract area with that of the Helvellyn map extract area.*

As can be expected from their position on the west side of Britain the Outer Hebrides display much evidence of the action of marine erosion. Wave-cut platforms, stacks, caves, cliffs and rocky promontaries are all features that are visible on the Isle of Lewis. They can be seen on the map extract and they occur on a grander scale in the extreme west of the island.

Sheltered inlets, with sandy beaches composed of material that has been swept up from a shallow sea, are common on the west coast as well. The sand is almost one hundred per cent calcareous and consists of small pieces of cockles and other shells. Often this sand is blown a short distance inland where it fills in lagoons and forms machair. Although machair is more widespread in the islands to the South, especially in North and South Uist, it does occur in Lewis.

Both west and east coasts are marked by broad inlets called lochs. These result from the submergence of pre-existing river or glacial valleys. Those in the east are more fiord-like, particularly where the land rises more in height.

**Human Activity:** In the Outer Hebrides it has not been possible to rely on the traditional occupations – farming, fishing and tweed making – for sufficient employment to guarantee that all the islanders will enjoy a satisfactory standard of living. For this reason many have left the islands and migrated to other parts of Scotland, or even further afield to countries such as Canada and New Zealand. Of those that remain today, almost half the work force is employed in professional and administrative services, which means a heavy dependence on the public sector. Only a tenth of the jobs are in farming and fishing, but their impact on the landscape is as significant as always.

**Crofting:** Over half the population of Lewis lives in coastal settlements known as townships. A township consists of a number of crofts that are usually strung out beside the main road. A croft is a small piece of land, between 0·5 and 2 hectares, upon which the crofter has the right to build a house. For this land a low rent is paid. The crofter also has access to common grazing land and to certain peat banks for fuel.

Some crops are grown on the machair land, but the majority of the crofter's income comes from the sale of livestock. Beef cattle and sheep are sold at the end of the summer, either for slaughter or for fattening on the mainland as stores. In the short growing season it is difficult for the crofter to produce enough winter fodder for the livestock, so this determines how many cattle and sheep are sold.

Crofting has always been a part-time activity, with a single croft unable to support a family. So crofts are often sub-let or abandoned. In order to promote the development of an area, where agriculture on its own is unlikely to provide the population with a reasonable standard of living, an Integrated Development Programme (IDP) has been set up in the Outer Hebrides. Funded by the Common Agricultural Policy and other EEC agencies the IDP aims to help crofters improve their working and living conditions.

A primary aim is to improve the quality of grazing and the amount of winter fodder that can be produced. Grants are available for land improvement that involves reseeding, drainage, fencing, lime and fertiliser application, creating access tracks, providing water supplies and shelter belts. Further grants are available to encourage an improvement in the quality of livestock. The success of the Programme will depend upon the ability of the crofter to make use of the financial assistance that is available.

**Fishing:** With many sheltered lochs along an indented coastline it is not surprising to find the fishing industry well established here. In the past fishing has often been associated with crofting and weaving as a way of life for many islanders. In recent years the trend has been to develop a full-time fleet that is able to compete against trawlers from other countries. Most fishing is done inshore for herring, mackerel, lobster and prawns, although depletion of herring stocks has resulted in temporary bans imposed on herring catches.

The Highlands and Islands Development Board, an agency established by the government to create employment in North and West Scotland, has made money available for the purchase of new fishing vessels. At Breasclete it built a new pier to cope with larger sized fishing vessels and the Board has helped install automated line fishing equipment and develop more fish processing plants. Fish farming is being encouraged in the many small lochs, particularly for trout, salmon and shell fish.

**Conservation:** Land improvement, the building of new access roads and the prospect that tourism will increase further may all lead to a conflict between development and conservation. The islands of the Outer Hebrides are of international importance for wild life. Conservationists suggest that the draining and re-seeding of the machair would destroy this environment and that if lochs are to be occupied by fish farms then the diving birds will be affected.

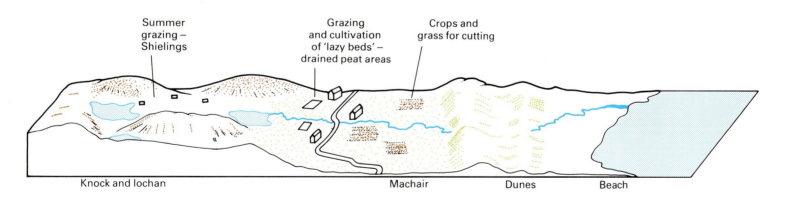

FIGURE 51

*Photograph Highlands and Islands Development Board*

## QUESTIONS ON THE ISLE OF LEWIS MAP

**Exercise A**
1. Give the six figure reference of (a) a pier (b) a beacon.
2. Study the islands of Orasay (218320) and Keava (198350). Then describe two ways in which they are similar and two ways in which they differ.
3. Locate the highest peak shown on the map extract, then give the name, height and six figure grid reference of this peak. What symbol marks the highest point?
4. What do the symbols at 238302 and 214295 have in common?

**Photograph Question**
(a) Study the photograph of Loch Roag and then state in what direction the camera was pointing when the photograph was taken.
(b) Draw a sketch to represent the area shown in the photograph. Use the map to help you label important features of the landscape.

**Exercise B**
1. Make a comparison between the area to the North East of Breasclete and that to the South East in respect of their relief and drainage. The minor road from Breasclete (218355) separates the two areas.
2. From evidence on the map and photograph suggest why a settlement developed at Callanish.
3. Consider the area to the west of the B8059. What evidence is there of rural depopulation? Suggest reasons for the area having so few settlements.
4. Describe the nature and course of the B8011 from the junction at 238318 to 209290.

**Essay Work**
Describe the features of a coast of submergence. Illustrate your answer with examples taken from anywhere in the British Isles, but you should include some examples from the North West of Scotland and South West Wales. Use diagrams where you can.

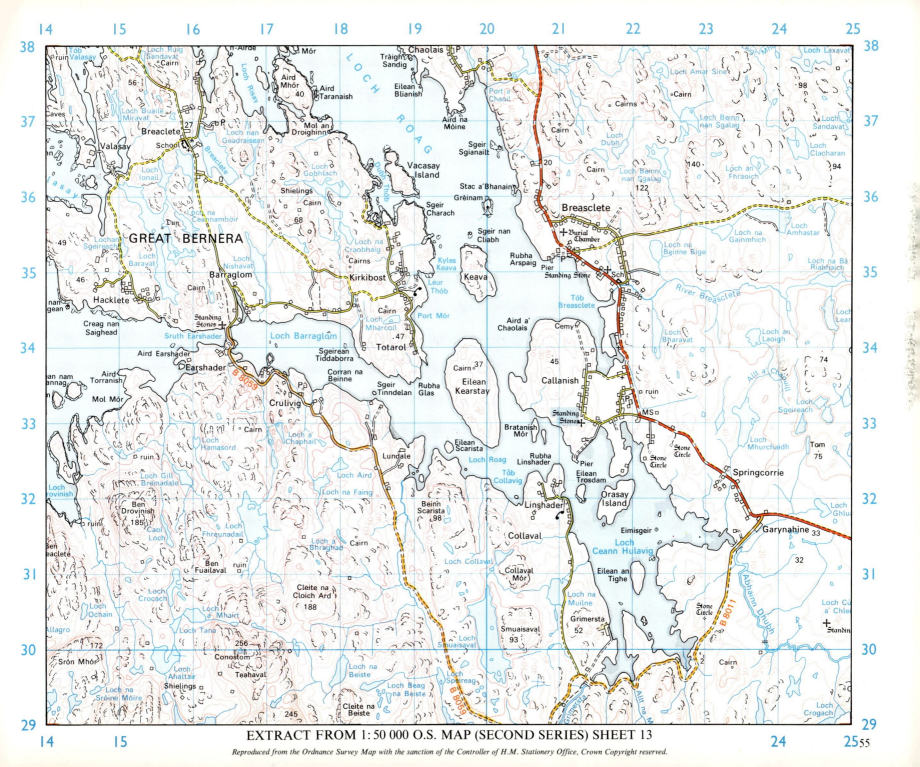
EXTRACT FROM 1:50 000 O.S. MAP (SECOND SERIES) SHEET 13
Reproduced from the Ordnance Survey Map with the sanction of the Controller of H.M. Stationery Office, Crown Copyright reserved.

# LINCOLN – AN URBAN STUDY

**The Development of Towns:** Most of us are familiar with a town. Eighty per cent of us live in or on the outskirts of one whilst the remainder must depend upon a nearby town for at least some essential services. Yet this degree of urbanisation is a relatively recent development; in 1770 only 20 per cent of the population were town dwellers.

A variety of factors encouraged the growth of large settlements. A town may have served as a focal point for the surrounding rural area, providing market facilities. To this function may have been added the administration of the region, as in the case of Taunton and Guildford. Although many towns established even as long ago as Roman times have maintained their importance to the present day (Lincoln being an example), the greatest urban development began as a result of the early stages of the Industrial Revolution in the eighteenth and nineteenth centuries. The development of railways saw the emergence of route centres, such as Crewe. In recent years planners have created new towns to offset the congestion present in some large metropolitan areas.

**The Structure of Towns:** Apart from the 'new towns', most urban areas have very similar land use patterns. Usually at the centre, where there is maximum accessibility, the shops and commercial enterprises of the town are to be found; this is the Central Business District (CBD). These concerns can afford the high rates and rents which this focal position demands. Then there are zones of industry and residential areas of varying quality.

In figure 51 we can see that industry is equally attracted to a central position as it requires easy access to supplies. Adjacent to it are the low cost residential areas for industrial workers, who need to keep the cost of their journey to work at a minimum. Those who can afford the time and money choose to live in the higher cost residences on the outskirts of the town. However, this concentric model does not allow for development. Industry and housing need room to expand; people are more mobile today and are able to afford the cost of commuting long distances to their place of work. The urban land use model, figure 53, allows for growth. Shops and offices still retain their central position but industry and housing advance outwards in sectors.

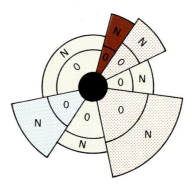

FIGURE 53
**Structure of towns – sector pattern. The letter O refers to older development, and N to more recent development.**

Not all commercial and industrial activities seek a central location in an urban complex. Dock industries have their locations fixed naturally, whilst industries such as large integrated steelworks need extensive areas of low cost land with accessibility to raw materials. Most significant is the congestion in urban areas, producing a resulting movement away from the centre. Whilst the CBD will remain, some concerns may develop on the outskirts of the town (figure 54).

*Exercise:* Consider your own town, or the town that serves the area you live in. Which model seems best to fit the land use there?

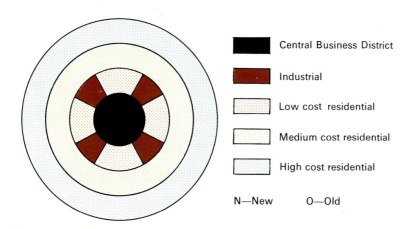

FIGURE 52
**Structure of towns – concentric pattern**

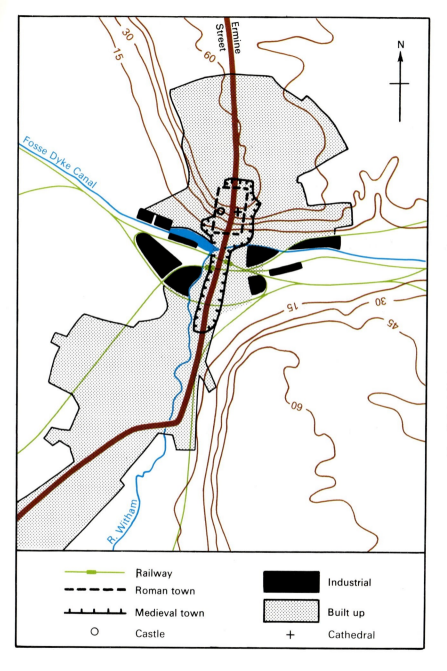

FIGURE 54
**Structure of Lincoln**

**The Development of Lincoln:** Lincoln is the leading administrative, cultural, industrial, retail, and market centre of Lincolnshire. It has a long history, and four periods of development stand out as being of special importance.

(1) The Roman Era. The Romans settled on the elevated site overlooking the Witham gap; this position on the Jurassic escarpment gave some protection from attack as well as commanding an important routeway through the gap. As Lindum Colonae this became one of the principal towns of the country as well as a market centre handling produce from those parts of fens which even then had been reclaimed.

(2) The Danish Period. Lincoln was one of the five fortified boroughs of the East Midlands and its trading links extended as far as Scandinavia.

(3) Medieval Period. The city prospered in the trade of wool until the end of the fourteenth century.

(4) Railway Era. Improved communications, with the opening of the Midland Railway from Lincoln to Nottingham in 1846, led to the growth of industry. This industry concentrated on the floor of the Witham, closely following the railways and the canal. Most of the original industries were set up to assist in the mechanisation of agriculture. Firms produced ploughs, threshing machines, steam engines, and pumps for drainage. Although engineering is still important today, industry has diversified and includes the manufacture of plastics, foodstuffs, paper, clothing, and electrical goods.

*Exercise:* Construct a land use map for Lincoln. Use the information on figure 52, the Ordnance Survey extract and the photograph.
(a) From the O.S. extract trace the railways, river, and canal.
(b) Use figure 54 to help you mark in the industrial area.
(c) Estimate the location and extent of the CBD; remember this is near the centre and near many lines of communication. Shade it in and label it.
(d) Locate the position of varying cost residential areas. Only informed guesses can be made using the extract, studying the road patterns (refer to page 21), and looking at the photograph which illustrates one section of the town. The following hints may help: terraced housing near industry in the low cost area; post-1950 housing estate is new middle-cost area; 1930–1950 housing areas are old middle-cost sectors.

Having completed the map, can you recognise any obvious correlations between the land use map of Lincoln that you have produced and the models on page 56?

*Photograph Aerofilms Ltd*

## QUESTIONS ON THE LINCOLN MAP

**Exercise A**
1. For each of the following grid references, substitute the features represented on the map.
    'You can easily travel to Lincoln by 023745, 020720 or even 938732. Relax there and enjoy visits to 978718 and 975718 on the original town site. If you would prefer to be more active then go to the 977696 or to the swimming pool in the suburb of 965697. However, for a little excitement 945692 is the place to go!'
2. Compare the relief in square 9471 with that of square 9767.
3. Direct a stranger to the information centre from the railway station at 973708. Include instructions about distance, turnings, gradient and any notable buildings or features that would be seen by the stranger on the journey.
4. Draw and label FIVE different symbols which occur along the railway line from 022700 to 040673.

**Practical Work**
Draw a topological graph to represent the following towns and mainroads: A15, A16, A17 and A18; Boston, Grimsby, Sleaford, Newark-on-Trent, Lincoln and Brigg. A road atlas will give you the information you need to construct the graph. Use the notes on page 13 to help you answer these questions.
   (i) Which is (are) the most accessible town(s)?
   (ii) Calculate the connectivity of the network.
   (iii) Add the links from Newark-on-Trent to Lincoln and Lincoln to Grimsby to represent the A46. What difference do these links make to
      (a) the connectivity of the network?
      (b) the most accessible town?
   (iv) Use the information to help explain why Lincoln is the county town.

**Exercise B**
1. (a) Draw an annotated sketch section from 950670 to 020670.
   (b) Using the section, comment on the relief and drainage of the region.
2. Describe the site of Lincoln.
3. Compare the pattern of settlement of the village of Canwick with that of Branston.
4. Locate the power station in square 9871 and suggest why that site was chosen.

**Research Work**
Make a study of the pattern of Roman roads which are followed by the roads of today, and of Roman towns which remain the site of present-day settlement. Draw a map to illustrate a short account on this subject.

**Photograph Question**
List FIVE distinctive features which are shown both on the photograph and the map extract.

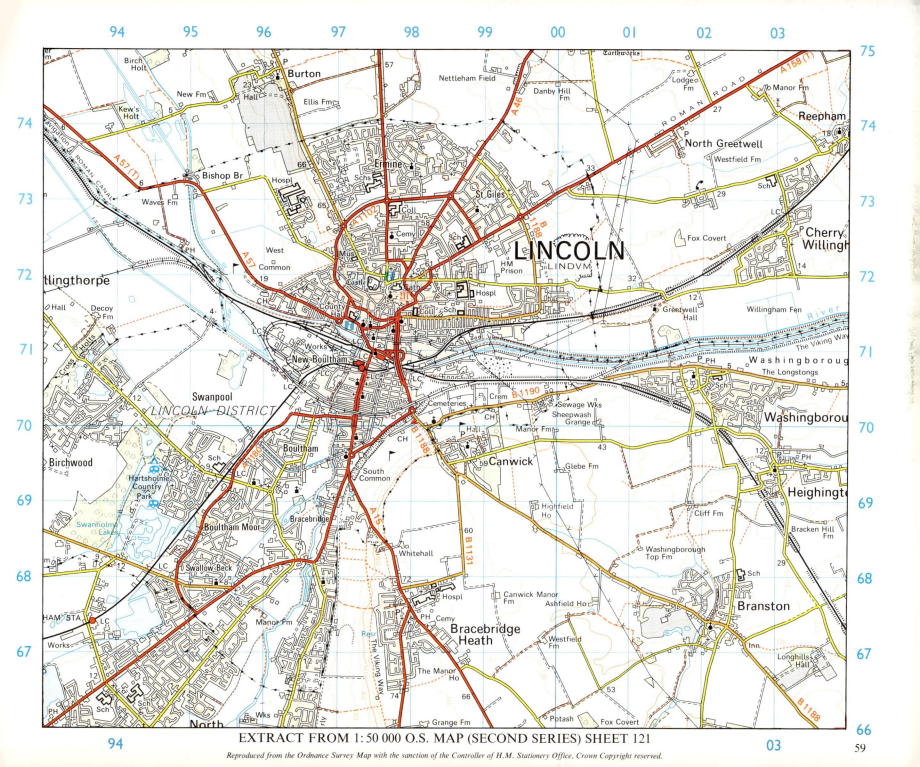

EXTRACT FROM 1:50 000 O.S. MAP (SECOND SERIES) SHEET 121

*Reproduced from the Ordnance Survey Map with the sanction of the Controller of H.M. Stationery Office, Crown Copyright reserved.*

# CORBY – A NEW TOWN

**New Towns:** Concern for the increasing congestion in our cities led to the 1946 New Towns Act. The aim was to develop self-contained communities which were spaciously planned, in contrast to the haphazard and cramped development of many towns since the beginning of the industrial period. It was felt that residential and industrial areas should be separate, industry served by efficient communication networks, and the residents provided with a high standard of amenities. Altogether the town ought to be an attractive environment in which to live and work.

New towns are established for one or a combination of the reasons listed below:
(1) To relieve overcrowding in a nearby congested urban area.
(2) To promote new industry in an area.
(3) To provide housing and amenities which are sufficiently attractive that present industry, which might otherwise stagnate, is able to expand.

Not all New Towns are being built in open country. Some are being developed from existing towns and in many cases the old town centre has served as a nucleus for the New Town.

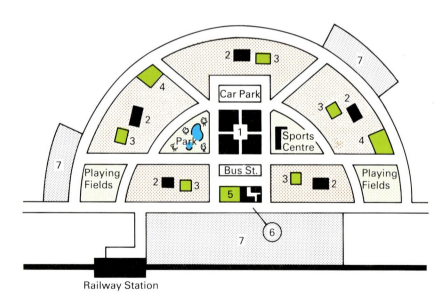

FIGURE 55
**Plan of a new town**

*Key:* (1) Shopping precinct – the main service area; pedestrians only; offices above theatre and dance hall. (2) Neighbourhood service centre in residential sector. (3) Junior/infant school. (4) Secondary school. (5) Technical college. (6) Library, police and fire stations, magistrates' court, and local council offices. (7) Industrial areas.

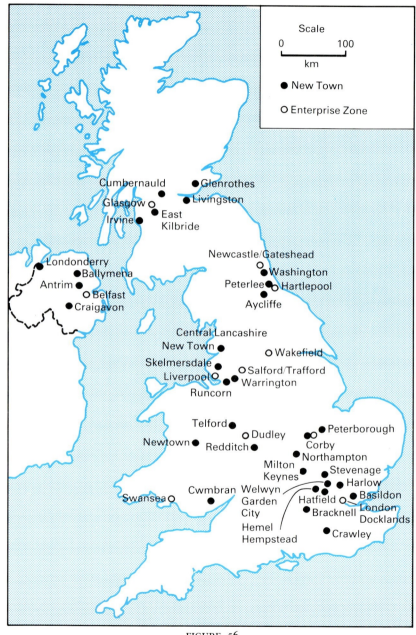

FIGURE 56
**New towns in Britain**

**Corby:** Up until the early 1930s Corby was a small Northamptonshire village occupied by some 1 500 people. The stone-built houses that clustered around the church belonged mostly to agricultural workers. A blast furnace, that smelted iron ore obtained from the underlying Jurassic rocks, was the only evidence of industrial activity.

The subsequent development of an integrated steel works at Corby created an enormous increase in the size of the population. Unemployment in Britain, particularly in central Scotland, caused a massive wave of migration to Corby as people came in search of a job. However, the village did not have the houses, shops nor recreational facilities to cope with such a rapid expansion. This pressure on the settlement's resources led to Corby being designated a New Town in 1950.

Whilst it was expected that the steel industry would continue as the main employer, smaller firms were attracted to the purpose-built industrial estates. Many of these firms were engaged in distributing goods or in making light industrial products. They tended to be employers of women, and this created a desirable balance as most of the steel workers were men.

In 1980 economic disaster struck at Corby when the steel works closed. (No longer was the site on low grade iron ore deposits an economic proposition.) There was an urgent need to attract new firms to the town to provide jobs for the 5 500 people who had been made redundant. Immediately Corby was given Development Area status, which enabled grants to be made available for industrialists. These benefits were considerably improved in 1981 when three selected industrial sites were designated the Corby Enterprise Zone. Industries that develop on these sites enjoy fewer planning restrictions, more tax concessions and fewer outlays, as local authority rates do not have to be paid.

Most of the firms that are taking advantage of these financial benefits are small. Although the tube making section of the steelworks remains and continues to be a major employer, a more diversified employment structure has resulted. Oxford University Press has established a distribution base at Corby and other distributive trades are being attracted to the town. With a population of 30 million people living within a 160 kilometre radius of the town, this alone might be an incentive for industry to locate there, besides the efforts that have been made to improve the accessibility of the town. *Why would an M1–A1 link road benefit Corby?*

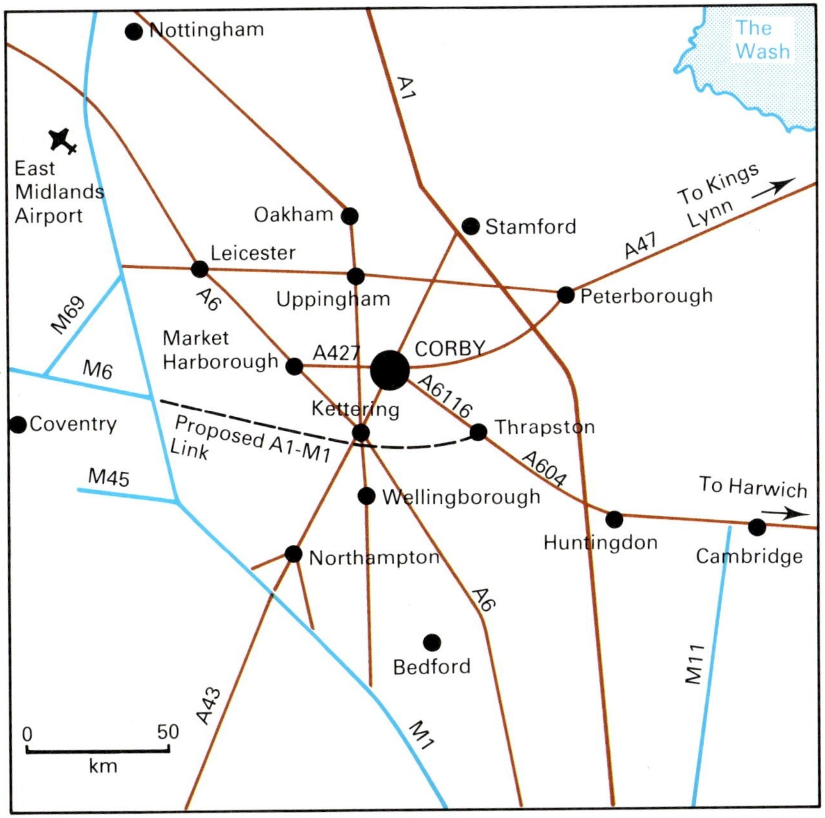

FIGURE 57
**Corby and the local main road network**

## QUESTIONS ON THE CORBY MAP

**Exercise A**
1. Write out the paragraph below, omitting the incorrect words.
   'From my house at 869937 situated next to Caldecott/Bringhurst church, I have only a short drive to work along the A6003. At the junction of this road with the B672 I turn south/north, continuing through the village/town. I am careful of the bend immediately before/after the Welland/Eye Brook bridge. Then I proceed over the flood plain to the village of Rockingham/Cottingham. Here the road falls/climbs sharply. At a Y junction I leave the A6003 and, travelling along the A6116, I pass the cemetery/quarry on my right and take the third turning on the left to my factory.'
2. How far has the traveller journeyed altogether in the question above?
3. Fit one of the given characteristics to each of these settlements:
   (i) Caldecott, (ii) Rockingham, (iii) Gretton, (iv) Bringhurst; (a) Dry point site, (b) Linear shape, (c) Wet point site, (d) Has two churches.
4. Why is the district boundary in the lake at 8594 so winding? Identify the man-made feature at 855943; this might help to explain the first part of the question.

**Exercise B**
1. Draw a plan of Corby on the scale used in the map extract. On this plan show the main lines of communication, the residential areas, the industrial zones, and the town centre.
2. Comment on the situation of the industrial and residential areas which you have shown on your plan.
3. Explain how the pattern of railways and spoil heaps helps to build up a picture of past industrial activities in the area.
4. Describe the course and valley of the River Welland. What stage has it reached?

**Essay Work**
Give reasons for the distribution of the Enterprise Zones shown on figure 56.

**Photograph Question**
1. What advantages has a shopping precinct as shown in the top photograph over a traditional 'High Street' shopping area?
2. What advantages has the layout of the industrial estate shown in the bottom photograph over an inner-city industrial zone?

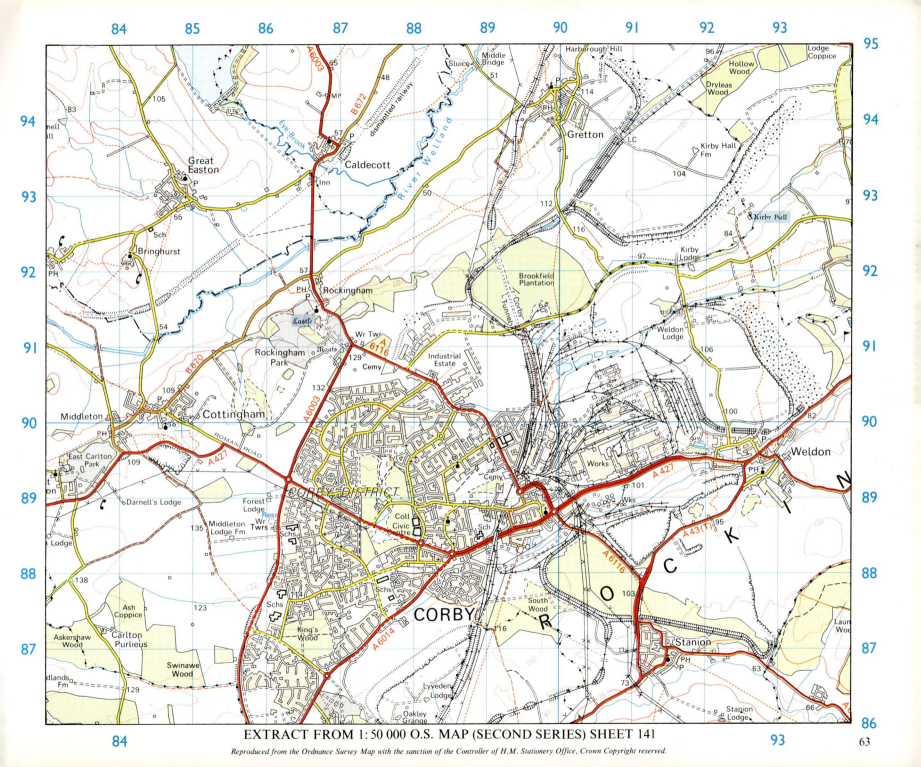

EXTRACT FROM 1:50 000 O.S. MAP (SECOND SERIES) SHEET 141

Reproduced from the Ordnance Survey Map with the sanction of the Controller of H.M. Stationery Office, Crown Copyright reserved.

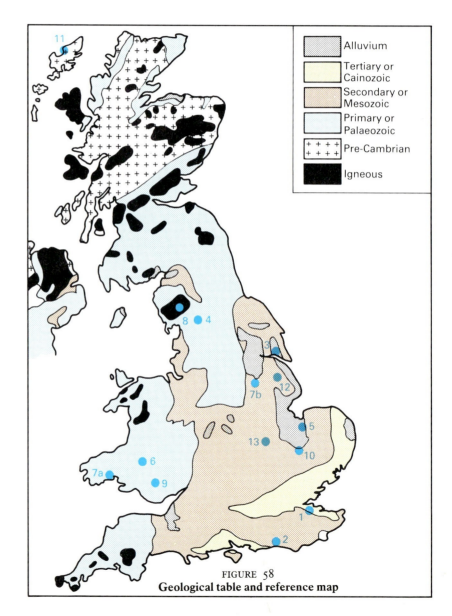

FIGURE 58
**Geological table and reference map**